DRILLING

Tools and programme management

JOHN S. HARTLEY
John S. Hartley & Associates, Mission Beach, Qld, Australia

A.A.BALKEMA/ROTTERDAM/BROOKFIELD/1994

Published by
A.A. Balkema, P.O. Box 1675, 3000 BR Rotterdam, Netherlands
A.A. Balkema Publishers, Old Post Road, Brookfield, VT 05036, USA

ISBN 90 5410 159 8 hardbound edition
ISBN 90 5410 160 1 student paper edition

©1994 A.A. Balkema, Rotterdam
Printed in the Netherlands

Contents

Acknowledgements

There were a great number of people who helped me with this book, and if in the following paragraphs I miss someone, please remember my memory is fading.

Firstly I thank those drillers who, by their knowledge and example of drilling practice, showed me the way. These were people like Allan Rowe, Brian 'Moscow' Forgan, and Tom Geheraty.

There were the people who helped me gather data to put the story together such as Duncan Maconachy, Simon Fitzgerald and Ray Cottrell of Longyear Australia, Ron Edgar of Leanda Drilling, Allan Rowe of Rowe Enterprises, and Roger Cooper who did a library search.

Then there were the draft readers, correctors, and suggestors such as Collin Kendall, Allan Rowe, Ron Edgar, Duncan Maconachy, Tony Alston, Grant Williamson, my son Robert, and more.

In particular I would like to thank Grant Williamson for the great effort in the many hours of drawing and redrawing most of the figures, and Hugo Becerra for work on the tables. I also thank the Australian Drilling Industry Training Committee Ltd (ADITC) for permission to use or adapt for use, from the *Australian Drillers Guide* those figures as indicated (39 figures). I also thank Longyear Australia for their permission to likewise use or adapt figures and tables as indicated (10 figures and 2 tables). I thank Sandvik Australia, R. Bluck, Cummings, Schunnesson, De Beers, and Pulisch for their contributions as noted.

I should be remiss if I did not acknowledge my previous fellow workers with Penarroya Australia Pty.Ltd. who first asked me to write a report on how I managed the drilling at Thalanga, so it could be used as a company text throughout the world wide operations of the parent company Imetal S.N. This was the starting point of writing.

Finally I would thank my wife, Dorothy, for her practical help with the wordprocessor, and for putting up with the mess I often created.

CHAPTER 1

Introduction

The objective of this book is to provide practical instruction for the geologist or engineer on the many facets of drilling and in particular diamond drilling. The book also makes useful reading for the drill operator so that he understands how his tools can be put to work most effectively.

This book is written by an Australian whose experience has principally been in Australasia. The description of methods and tools reflects this and the author expects that there will have been parallel development in other parts of the world which have not been specifically cited here.

From the *Australian Drillers Guide (1985)* some questions the geologist/engineer should ask are presented:

1. Should I make sure the driller is told the purpose of the holes he's drilling?

2. Should I tell the driller how much I'm depending on him to provide as accurate information as possible and how this information may affect future decisions regarding the project?

3. Should I let the driller know where the project is heading by telling him the project objectives so that I get his co-operation and understanding?

4. Should I get to understand more about problems the driller faces so that we can communicate more fully?

5. Should I get a greater understanding of the rig's capabilities so that I can appreciate more readily the contribution of the driller and his machine?

6. Should I discuss with the driller the necessity for balancing expenditure on drilling with expenditure in related areas to ensure maximum benefit to both parties?

7. Should I be able to discern easily the differences between the competent driller and the one who is slack?

8. Should I acknowledge and reward the driller for the extra care he takes to provide additional information?

9. Should I include the driller as part of the team and help him in matters to do with interpreting data or predicting the sort of formation he's likely to drill?'

These questions should be continuously asked throughout one's career.

The following chapters on types of drill, rod strings, and sampling and cutting

1

tools are meant as background for the uninitiated, such as students, new graduates or new drill hands.

The chapters that follow these, are ones which are of relevance to all and contain checklists and tables which will be of lasting value.

Drilling is important to geologists and drillers.

Drilling makes up 'at least' half of a well organised exploration company's budget and forms a large part of mine planning and production costs. It is virtually the only way to prove up a natural resource of mineral, oil, coal, or water. Successful drilling will be a win for geologists and also for the drillers whose business is drilling.

Why do we drill?

Drilling is to make holes in the ground for a variety of reasons such as:
- To sample for Chemistry, Rock Type, Structure, Geotechnics;
- To enable water or oil to be sampled, pumped;
- To pump grout etc. into the ground;
- To provide cable access;
- To enable blasting of rock.

How do we drill?

This variety of purposes will impose different methods and different importances to the drilling parameters, such as recovery of core or cuttings, accuracy of end of hole location or straightness of hole, for example.

There are often perceived differences in objectives by the various participants once drilling commences.
- The driller wants a hole as quick as possible;
- The contractor wants a hole as cheap as possible;
- The geologist wants a hole as accurate as possible.

Each of the participants should take an active role in the education of the others to enable all three objectives to be realised.

Over the past 30 years that I have been involved with drilling I have seen a merging of these differences as tools became available which enabled the driller to more easily accommodate the other parties' objectives.

These tools have also enabled drilling costs to be contained. In 1960 for example, the base cost of diamond drilling one metre (*A* size core) in Australia, varied from 35% to 50% of the average weekly earnings of about $A40 ($US28). In 1990, the base cost of diamond drilling one metre (*N* size core) varied from 12% to 20% of average weekly earnings of about $A500 ($US350).

This progress has been made and will continue to be made by an exchange of knowledge by the participants. Such an exchange can be made:
- On site: Question and answer, suggestion;
- At training sessions: Industry training courses, University / College courses, in house, conferences, own Association, industry related.

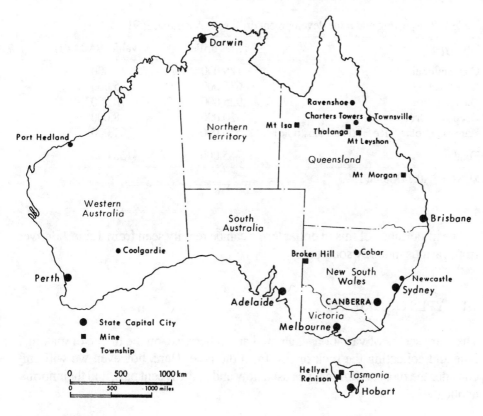

Locality map.

As drilling has become more sophisticated, (i.e. you need more in your tool box than the roll of fencing wire, which a particular contractor was reported to have as standard equipment), so the drillers have had to be smarter, and responsible for more than just hole making. To do this effectively, they have to be interested in improving their knowledge and skills other than by trial and error.

Geologists have, in general, not kept pace with drilling developments, and are rarely in possession of enough knowledge to make the selection of drilling techniques or tactics.

If then, we can get better communication between drillers and geologists, it should be within the framework of drillers teaching geologists what they do, how they do it, and why. Geologists must explain why the holes are being drilled and what may result, and what these results will mean. In order to fully appreciate the way rocktypes and ground conditions affect drilling, there needs to be a continuing dialogue and perhaps experiments.

It is hoped that by this brief exposure to the complete scene, geologists and drillers may be better able to assist each other to obtain the desired result as professionally and economically as possible.

Table 1. Prospecting and mine development drilling in Australia, 1991.

Drill type	Metres drilled	Value ($A000's)
Conventional	125,000	15,000
Underground	635,000	35,000
Dual purpose	925,000	70,000
Deep Dir. Drill	20,000	8,000
Hammer/Rotary (RAB) (excl. blasthole)	1,550,000	31,000
Total	3,255,000	$155,000

$US1 = $A0.7 (March 1993)

The importance of this in dollar terms can be readily seen from Table 1 derived from various industry sources.

1.1 TYPES OF DRILL

There are scores of variations, patented and otherwise, on the theme of making a hole and collecting the rock or soil from the hole. Here, however, we will only consider the most common types used now and in the recent past, and their normal application.

1.1.1 *Wash drill*

This is probably the simplest of all hole making devices as there are no mechanical parts. A piece of tube (casing) with a hardened shoe bit is pushed into a soft formation with a water or air flow jetted at the face (Figure 1). This type of drilling is usually an accessory to other drill rigs for use in very soft sluiceable ground, from which a disturbed sample is satisfactory.

1.1.2 *Hollow rod drill*

This is really a variation of the cable tool drill (discussed later). This time instead of cable, rods are used down the hole with a valved bailing bucket behind the bit (Figure 2). These are only used in unconsolidated material such as alluvials. With casing moving forward at the same time, they give quite good uncontaminated samples although there is no precise depth definition.

1.1.3 *Auger drill*

The principal component is a slowly rotating rod, with feed by mechanical or hydraulic pressure. The material from the hole is forced upward by the wedging

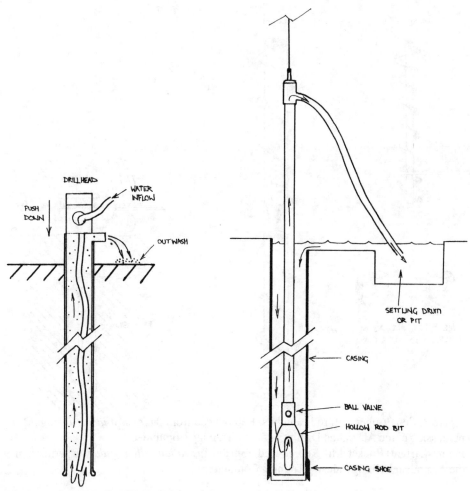

Figure 1. (left) Wash drill. Reproduced from the *Australian drillers guide* by permission of the Australian Drilling Industry Training Committee.
Figure 2. (right) Hollow rod drill. Reproduced from the *Australian drillers guide* by permission of the Australian Drilling Industry Trading Committee.

action of the bit tips, and in most cases, by a spiral ledge fixed to the exterior of the rods. Some models have hollow rods which can simultaneously take an undisturbed sample (Figure 3).

The rig size can vary from a small hand-held chainsaw adaption to drill 25 mm holes to 0.6 m, and similar small air driven augers for coal sampling, up to truck-mounted rigs which will drill up to 400 mm diameter holes to 40 m deep in the smaller diameters. These rigs are mainly used for post holes and sampling soil, clay and shallow weathered rock, in situations where a disturbed sample is acceptable. These rigs are not suitable in unconsolidated alluvials or very wet conditions or for hard rock.

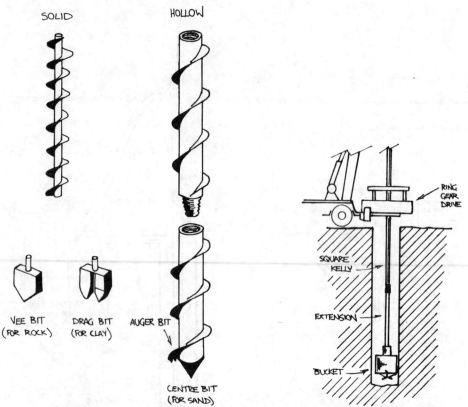

Figure 3. (left) Auger drill rods and bits. Reproduced from the *Australian drillers guide* by permission of the Australian Drilling Industry Training Committee.
Figure 4. (right) Bucket drill. Reproduced from the *Australian drillers guide* by permission of the Australian Drilling Industry Training Committee.

1.1.4 *Bucket drill*

These are a type of auger without the spiral, the material being forced up inside a bucket until it is full.

These rigs range in size from small hand drills such as post hole diggers capable of drilling up to 200 mm diameter holes to 10 m depth, to large truck mounted rigs such as a Caldweld which will drill up to 1.83 m diameter holes to about 60 m depth. Rotation is by a square section shaft through a kelly and feed is by gravity, with retrieval of cuttings by winching up the bucket (Figure 4) and the drive shaft remaining down the hole.

The small hand-held types are mainly used for post holes and geochemical soil sampling. The larger rig is used for water wells, shallow shafts (e.g. in opal mining), building pylon foundation holes, and bulk sampling (e.g. alluvials, coal, saprolite, gossan). Most rock which can by ripped by a dozer can be drilled in this way. Casing can readily be used, and can be advanced with the bit.

1.1.5 *Rotary drill*

This method uses a slowly rotating rod string on the end of which a bit scrapes (blades), or chips by point pressure (roller). The flushing of cuttings may be by air (*RAB* – Rotary Air Blast), water or mud. Circulation may be normal (i.e. cuttings coming up between the rod and hole wall or casing, or circulation may be reverse (*RC*) (i.e. cuttings returning by the interior of the rods). Air coring is a specialised variation of *RC* in which an annular bit, similar to a diamond drill bit, cuts soft, easily broken rock, which air blows through the centre of the rod to the surface.

Another specialised version is the now outmoded calyx drill, in which small chilled steel shot was fed to the face where a hardened steel bit rotated to give a milling action. This usually cut an annulus. The main use was for small shafts up to 2.4 m diameter and to depths of about 300 m. It was only used in very steep holes (to keep the shot evenly distributed at the face) and only in unfissured ground so that shot would not be lost. It is now common to refer to a calyx bit as one which cuts core but has no hard face or diamonds, just teeth cut into the rod.

As usual there is a wide variety of rig capacities. These fall broadly into three categories according to their principal use. A typical rod string arrangement is shown in Figure 5. It should be noted that the same basic machine and rod string are used for downhole hammer drilling described later.

Light weight

These are mostly for holes 90-120 mm in diameter, to about 80 metres in depth. The main use is geochemical sampling, and shallow, low volume water bores. These are mostly conventional circulation air flushing drills. This *RAB* drilling can often provide sufficiently reliable samples to at least provide indicated resource estimates. Providing the hole walls are reasonably stable it just requires care in sample collection.

Medium weight

These drill mostly in the range of 120-240 mm diameter and to depths of about 1500 m. These rigs are principally used for water bores and air coring, and are the work-horse of seismic drilling operations.

Heavy weight

These are the traditional oil drilling rigs which can drill holes up to 1 m diameter and to depths of about 8500 m (of course it is not maximum diameter all the way).

1.1.6 *Percussion drills*

These are based on having a slowly rotating rock chisel hitting the rock face, with the cuttings (chips) being variously removed. The method of application of force to the chisel allows a classification of these drills.

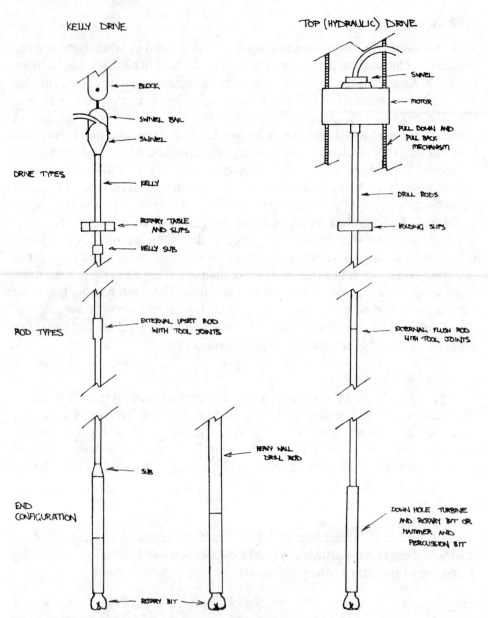

Figure 5. Rotary drill rod string. Reproduced from the *Australian drillers guide* by permission of the Australian Drilling Industry Training Committee.

Cable tool drill

This is one of the survivors of primitive drilling devices and is also known as a mud puncher. In the simplest configuration a large heavy chisel (about 0.5 tonne) is attached to a wire cable (Figure 6), the other end of which is attached to a slip winch which repeatedly lifts the chisel and allows it to free fall (of the order of

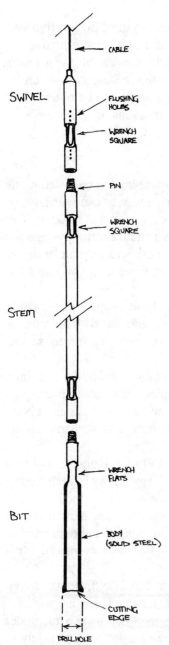

CABLE

SWIVEL

FLUSHING
HOLES

WRENCH
SQUARE

PIN

WRENCH
SQUARE

STEM

WRENCH
FLATS

BIT

BODY
(SOLID STEEL)

CUTTING
EDGE

DRILLHOLE
DIAMETER

Figure 6. Cable tool downhole assembly. Reproduced from the *Australian drillers guide* by permission of the Australian Drilling Industry Training Committee.

45-101 cm). Rotation is provided by twisting the cable back and forth by hand (the timing of which is an art). Removal of cuttings is done by bailing at periodic intervals with a valved bailing tube. Water to aid sample removal, if not naturally present, is just poured in. These holes must be almost vertical as gravity is the driving force.

These rigs commonly drill 200-400 mm holes to a depth of 300 m (but can reach 2000 m). Although quite slow, these rigs are still quite cost effective for deeper water bores, (i.e > 100 m) and they do produce straight holes. Another major use is for testing unconsolidated alluvials, where, because casing can be driven ahead of the bit if required, a very large clean sample can be obtained. A tube can be dropped instead of the chisel in soft formations and hence obtain relatively undisturbed and uncontaminated samples.

Uphole hammer drill

These vary from hand-held jackhammers up to somewhat larger variations mounted on a feed boom, which in turn is mounted on a trailer, tracked carriage, or truck. Compressed air fed to the hammer provides percussion and rotation. The rod string is small diameter of very tough steel, with a small central passage for flushing fluids, usually air or mist. As we see later, these small diameter dense rods are needed to impart surface generated percussive force to the bit down the hole.

With energy losses at each coupling, there are depth limitations. There is also a very big difference in the area of the input air passage compared with the annular area of exiting air with cuttings. The resultant lower velocity means severe limitations as the hole gets deeper.

The bit is either some variety of chisel or number of spherical buttons set on a near flat surface. The drill rods are externally coupled solid or flush coupled tubular. The rig is usually mounted either on a small two wheeled trailer, or is on an air motor driven, metal tracked carriage which may tow or carry an air compressor on board.

The capacity varies from 25 mm diameter up to 150 mm diameter, and to a depth of 150 m. However, they rarely drill more than 30 m as they have difficulty with damp clays and hence the watertable. These are mainly used for blast hole drilling, but they can be useful for first pass exploration holes particularly in rugged terrain when track mounted. Declined holes are possible, and directional accommodation is usually excellent, with underground versions commonly drilling holes inclined upward.

Useful small versions are the Cobra and Pionjar, hand-held, petrol driven jackhammers. Flushing 'air' is provided by exhaust gasses. These are used for 25 mm holes to break boulders, dig trenches, etc. to a depth of 6 m, under favourable conditions. Because of exhaust gasses they cannot be used safely in confined space, such as pits more than 2 m deep. There are also small electric drills with eccentric cam to provide impact, but these must have a separate flushing fluid source.

Modern hammers are powered by hydraulic fluids to provide impact and with electric motors to provide power to the hydraulic pump. Oil under pressure is a more efficient driving medium than air, as it is less compressible. Flushing air or water are sourced separately.

Downhole hammer drill

These machines have a similar hole-making principle to jackhammers, but in these the hammer is attached directly to the bit at the end of the rod string. The percussion is provided by the hammer down hole; the rotation is provided at the top of the rod string by a separate motor, usually hydraulic. Feed is also supplied from the top. The string is a much larger diameter pipe, rather than rods, to allow the additional drive air to be available to the hammer. This has a side benefit of providing a large amount of exhaust air. It is this availability of more air at the face which is one of the reasons that allow these rigs to penetrate much further than uphole hammers, and to more readily handle clayey material. The almost flush coupled rods provide more uniform clearance between rods and wall, and so give greater average return air velocity, so allowing much better hole flushing.

Downhole hammer drills make holes which vary from 100 mm to 650 mm in diameter, and drill to more than 1500 m. These rigs provide the quickest method of disturbed, but depth orientated, sampling of medium to hard rock especially from the watertable to 300 m. Being relatively inexpensive they make ideal precollars for diamond drill holes (about 30% of cost). With care, and with some simple accessories the deviation of the hole can be controlled to some extent. Because of less rod clearance there is much less deviation than in uphole hammer holes.

The integrity of the sample will largely depend on circulation method and fluids. For example reverse circulation with air is very good, but in wet holes there will be more contamination. This will be discussed further in Chapter 15.

1.1.7 *Diamond core drill*

These involve driving a fast rotating annular bit through rock and so obtaining a solid undisturbed sample. The bits are either diamond impregnated or surface set. Occasionally other materials such as tungsten may be used. In terms of sample weight or lineal metreage this is at least three times as expensive as any other drilling method, at similar depth.

The primary power source for the rig may be diesel (mostly on surface rigs), or air or electricity (for underground). Apart from the primary power, there is a classification of rigs based on where the drive power is applied. Rotation can be applied through an in-line motor (usually hydraulic), at the end of the rod string, called topdrive, or it can be applied through sets of gears to a chuck which fastens to the outside of the rods, called bottomdrive.

Feed with topdrive is more continuous, usually by a chain and sprocket driven by hydraulic motor. Feed for bottomdrive is now only by short hydraulic rams which means advance is more a batch process.

The apparatus is usually mounted on a frame. This includes the drill, mast and water circulation pump. This frame may be mounted on skids to be towed or pushed by dozers or such, or it may be fitted with jack up-legs (hydraulic rams), to allow travel on the back of a truck, but also to allow it to be easily placed on ground to drill. Alternatively it can be mounted on a truck or tracked device as a permanent configuration. For underground or helicopter supported operation, the components are often individualised, to enable quick manual separation into units of about 500 kg.

Although in special circumstances air can be used, water is the common circulating fluid, with cuttings washed up the hole between the rods and hole wall or casing. To the water, various additives are added to improve performance.

Core is collected at the bottom of the hole in a core barrel and brought to the surface periodically (more on this later). Core size can vary from 37 mm to 64 mm diameter in the commonly used configurations and depths to 5000 m are attainable. Table 2 shows the hole and core sizes which are relatively common.

It will be noted that there are a whole host of core sizes depending on the type of bit and barrel (core collecting tube) used. These are coded with internationally recognised letter symbols. This is an outline of how it works:

1. The initial letter i.e. *E, A, B, N, H, P, S*, etc. refers to the hole size.

2. The following letters indicate different rod (thread type) sizes e.g. *X* = old original, *W* = standard, *Q* = wireline, *J* = tapered thread.

Table 2. Common diamond drill core and hole sizes.

Size	Core diam. (mm)	Hole diam. (mm)
AQ, AQ-U	27.0	48.0
BQ, BQ-U, BQWL	36.5	60.0
NQ, NQ-U, NQWL	47.6	75.7
HQ, HQWL	63.5	96.0
PQ, PQWL	85.0	122.6
BQ-3	33.5	60.0
NQ-3	45.0	75.7
HQ-3	61.1	96.0
PQ-3	83.0	122.6
RWG	18.7	29.8
EWG, EWM, EWL, EX, EXM	21.5	37.7
AWG, AWM, AWL, AX, AXM	29.6	48.0
BWG, BWM, BWL, BX, BXM	42.0	59.9
NWG, NWM, NWL, NX, NXM	54.7	75.7
HWG	76.2	99.2
BQ2.32	38.6	58.9
NQ2	50.7	75.7
HQ3.18	66.2	93.5
LTK46	35.6	46.2
LTK56	45.2	56.3
CHD76	43.5	76.3
CHD101	63.5	101.3
CHD134	85.0	134.3

3. The next letter or number refers to barrel or bit type or rod thickness e.g. *T* = thin walled bit, *LC* = Variety of split inner tube, 3 = triple tube.

A more complete listing of possible core and hole sizes appears in Appendix 1.

A conventional standard core barrel requires all the rods to be removed from the hole before the core can be removed. In 1961 the wireline system was introduced. This enables an innertube with the core to be withdrawn through the rods on the end of a cable or wire. This is now universal equipment, with a conventional barrel only being used on special occasions.

Diamond core is difficult to obtain from unconsolidated material for two principal reasons:

– Large gravel pieces tend to either move whilst being cut; they remain whole or they jam in the bit so not allowing it to perform its usual coring function;

– Fine sand, whilst feeding into the barrel, will not be rigid enough to be retained by the core lifter.

1.1.8 *Hydro-percussive diamond core drill*

He Yizhang (1987), reported the development in China of a vibrating diamond bit (3000 impacts/min of 12.7J), using hydraulic pressure, through a double acting or spring valved hammer in the core barrel. This is said to be very good for badly fractured rock, where run length has been shown to increase from say 0.5-0.6 m to 2.0-3.0 m. By reducing feed pressure there should be less deviation as well as advantages such as increased penetration rate, longer bit life, and decreased costs. The reported improvements in performance are impressive, but, clearly, this is still experimental, though it would appear a likely path in the future.

1.1.9 *Multipurpose drills*

Multipurpose or universal rigs are, as their name implies, capable of rotary, downhole percussion, and diamond drilling. This has been made possible by the development of the topdrive configuration, which allows the slow revolutions and high torque required for rotary as well as the high speed required for diamond drilling. They are equipped to circulate water and mud or air. Although the same rod string can be used for rotary and hammer it is usual to have a separate diamond core string available. Should a reverse circulation string be present it can be used equally well for conventional circulation. The most important use for this type of drill is where core is not required until the target is reached, so reducing the average drilling cost of the hole. However a note of caution should be entered here; rotary and hammer holes are more difficult to control and the contractor may make significant charges to change rod strings. Look at the real average costs of the whole hole.

CHAPTER 2

Rod strings

There are a variety of rod types available, each with a particular application, as shown in Figure 7.

Firstly we can categorize rods by the manner in which cuttings are extracted.

– Mechanical;
– Fluid.

Secondly we can categorize rods by the manner in which the fluids are circulated.

– Standard circulation;
– Reverse circulation;
– Dual tube reverse circulation.

Standard circulation is where fluids pass down the centre of the rod to the face, and then the cuttings pass up between the rod and hole wall or casing.

Reverse circulation is where the fluids pass down outside the rods, and up with cuttings inside the rods.

Dual tube reverse circulation is where the fluids pass down to the face within the annular space between the drive rod and an inner light weight rod (attached to the drive rod). The fluid and cuttings then pass up the inner rod to the surface.

The everyday use of the term Reverse Circulation, or its commonly used abbreviation, *RC*, usually means dual tube reverse circulation, and from here on the same will apply, unless otherwise indicated.

The dual tube *RC* system was brought into use to enable downhole hammers to work with *RC* in that clean air was brought to the hammer first. The impetus for usage was brought about by the need for cheap uncontaminated sampling of deep saprolitic soils, which characterises many of the latter day West Australian, low grade gold deposits. Standard circulation often smears out the values as cuttings often adhere to the hole wall or on rod couplings, etc.

For many years these *RC* rod strings were used with a conventional downhole hammer, which meant that immediately above the hammer there was a sub (connection) which took the clean air from the annulus to the centre of the hammer, and then as the air and cuttings returned up around the outside of the hammer (i.e.not *RC*), and reached the sub, they were diverted to the inside tube.

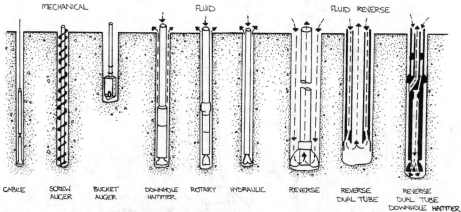

Figure 7. Drilling methods, rods and hole clearing. Reproduced from the *Australian drillers guide* by permission of the Australian Drilling Industry Training Committee.

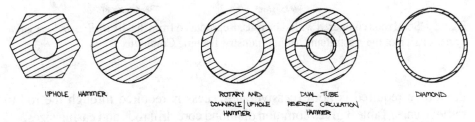

Figure 8. Drill rod cross sections.

This sub is known as a crossover sub. Effectively then, there was only standard circulation for approximately 1.5 m from the face.

Lately there has been the development of downhole hammers which allow the circulation to remain reverse right to the face.

The wall thickness in various strings may vary (Figure 8).

1. Small diameter, small orifice rods with thick walls are used for small uphole hammers to transmit percussive and rotational forces over short distances. Larger hammers have allowed the use of thick walled tube rods, in most cases with flush jointed rods without couplings. These are better for power transmission, return air flow, and straighter holes (Figure 9).

2. Large diameter thick walled rods are used for rotary and downhole hammers to enable greater torque and air at the bit.

3. Downhole hammers may use large diameter thick walled rods or a combination of thin walled rods for much of the hole but with thick walled (collar) rods at the bottom of the string to provide the weight behind the hammer but only moderate torque.

4. Moderate diameter and thin walled rods are used by diamond core drills as

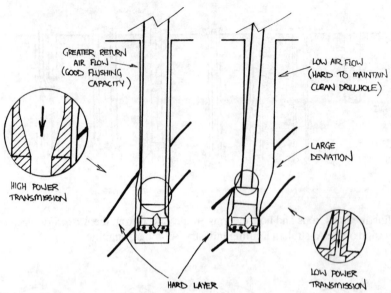

Figure 9. Solid rods compared with tube rods. Reproduced from the *Australian drillers guide* by permission of the Australian Drilling Industry Training Committee.

the torque required is not excessive and access is required through the rod to retrieve core. Table 3 gives common diamond core drill rods and casing sizes.

As an indication of the order of strength differences compare the following:
- 59.4 mm O.D. Rotary (86mm box) 490 kg/m torque;
- 55.6 mm O.D. *BQ* diamond 125 kg/m torque.

Despite the apparent rigidity of individual rods the complete rod string is quite flexible. Up to 10° deviation over 30 m is quite common for *BQ* in schists.

An example of considerable deviation is shown in Figure 10, which shows a hole at the *XYZ* mine, which because of two different regimes of anisotropic strength, has deviated 250 m south of its intended target.

Another example is at Thalanga Mine, Queensland, Australia, where 70° declined surface drill holes (*BQ*) commonly can deflect upwards 7° per 30 m, but up to 14° can be forced with excess thrust.

These are also examples of why downhole surveys should be done as the hole progresses.

The type of thread used on rods depends on their application. For example, in simple terms, uphole hammer rods have rope threads, downhole hammer/rotary rods have truncated *V*-section tapered threads, and diamond drill rods have square section box threads with a slight overall taper.

Rods will wear and fatigue in various ways for various reasons, but the three which a geologist should be aware of are:

1. Sharp bends in the hole followed by hard rocks will mean that each rod will

Table 3. Common diamond drill rod and casing sizes.

	O.D. (mm)	I.D. (mm)	Weight kg (3 m length)
Rods			
EW	34.9	30.9	13.8
AW	44.4	40.4	19.7
BW	54.0	50.0	18.8
NW	66.7	62.7	24.2
XR	25.4		
XRT	27.8	30.2	
E	33.3	21.4	
A	41.3	32.1	
B	48.4	35.7	
N	60.3	50.8	
AQ(HT)	44.5	34.9	14.0
BQ(HT)	55.6	46.0	18.0
NQ(HT)	69.9	60.3	23.4
HQ(HT)	88.9	77.8	34.4
LTK46	43.0	37.0	21.8
LTK56	53.0	45.8	33.2
Casing			
XRT	28.8	30.2	7.8
EX	46.4	41.3	8.0
AX	57.1	50.8	12.5
BX	73.0	62.7	26.3
N	88.9	77.8	34.4
H	114.3	100.0	
RW	36.5	30.1	
EW	46.0	38.4	
AW	57.1	48.4	17.0
BW	73.0	60.3	31.3
NW	88.9	76.2	38.5
HW	114.3	101.6	50.6
PW	139.7	127.0	64.4
SW	168.2	152.4	
UW	193.6	177.8	
ZW	219.0	203.2	

be in the bend for a longer time and the whipping action will fatigue to failure. Remedy is not to deviate or deflect holes too rapidly. If you want the hole to bend do it over a greater length. Deviation rates greater than 7° per 30 m may cause trouble. Where sharp bends do exist then the rods that pass this point should be very tightly coupled to avoid overflexing of the joints.

2. If feed pressure is too great and the rods are bent (Figure 11) the active length

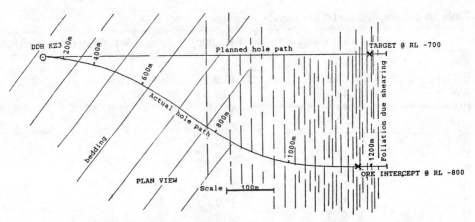

Figure 10. Deviated drillhole. XYZ Mine.

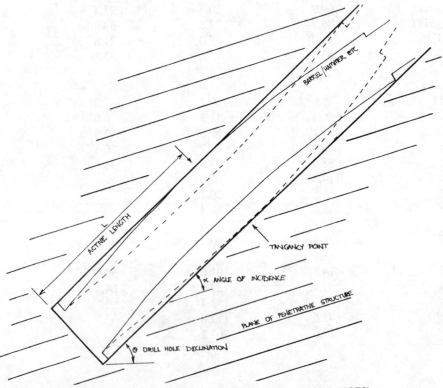

Figure 11. Rod and barrel bending under compression. From Bluck (1978).

is decreased and the hole will deviate more, resulting in (1) above, or belling of the rods.

3. Wear by abrasion will be greater in some rocks (e.g. friable quartzite) which will increase drilling costs.

CHAPTER 3

Hole-making tools

This chapter will introduce the various hole-making tools, such as bits, how they 'cut', and how the cuttings are removed. Further discussion on cutting removal will be found in Chapter 15.

Firstly consider all the physical actions which may be used to make a hole:

Crushing, Impacting, Tearing, Ploughing, Twisting, Grinding, Cutting, Shattering, Abrading, Sluicing, Eroding, Spalling, Stirring, Displacing, Excavating.

Most-hole making techniques are combinations of these. Later in this chapter, we shall examine the principal drilling methods and see which of these physical actions apply and how. Before doing that we will consider the principles of hole-making.

Having the correct thrust plays an important part in making sure that the various drilling methods work effectively (i.e. use the appropriate physical actions to their fullest extent).

Thrust may also be termed bit loading, pulldown, or down pressure. This is the force used to keep the bit on a fresh face, but in a hammer drill does not include the impacting force. Thrust can be either static (i.e. weight of rods and sometimes rig), or can be applied by mechanical, hydraulic, or even manual feed pressure.

In percussion drilling adequate thrust is needed to transmit the impacting forces generated by the hammer to the bit and hence the rock face. If the force is insufficient the bits bounce around, wearing but not cutting the rock, and the recoil not being dampened by adequate thrust will cause excess rod vibration, and therefore, more wear and hole caving problems. Increasing thrust will increase penetration until eventually, it will cause a stall. The load point for thrust is also important as heavier rods will be better than top hole force, due to the inherent elasticity of any rod string.

In rotary drilling, penetration speed is proportional to thrust, provided there is sufficient torque. Therefore the thrust used is related more to the size of the rig, the strength of the rods and the hardness of the rock being drilled (e.g. high pressure in soft clay would block the circulation). Overthrust will, before stalling, cause rod bending and consequent deviation. Underthrusting will cause excessive bit wear.

Diamond drilling relates to thrust in a very similar way to rotary drilling.

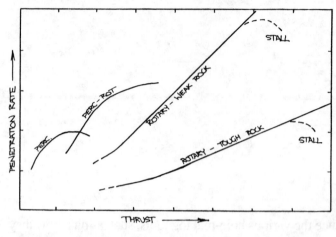

Figure 12. Effect of thrust on penetration rate. From McGregor (1967).

Figure 12 from McGregor (1967) shows the relationship between thrust and penetration rates for percussion (hammer) and rotary drilling. Here we see that percussion drilling requires little thrust, and other forms of drilling proceed faster if more thrust is applied. The stall situation is related to the excessive thrust where insufficient torque is available.

Bits, rod strings and the rig are more likely to be damaged when high thrust and rotation are attempted at the same time. Therefore rocktypes and drilling techniques requiring greater thrust are drilled with lower rotation speeds.

3.1 AUGER BITS

With a solid rod auger, the bit is a variety of blade bit, the vee bit being used for harder rock than the drag bit (see Figure 3). The hollow rod auger which may also obtain a core sample, has hardened steel or tungsten carbide inserts or attachments on the leading edge of the auger flights. A solid rod auger may use a similar device for soft rocks. The principal actions used are cutting and ploughing, with the soft clay and the like being pared off like a plane shaping wood.

3.2 CABLE DRILL BITS

There are two major species of bits which are in common use:

3.2.1 *Chisel bits*

These are very large single blades or star shaped. They are mostly hardened steel (hand forged) or hard faced by welding (Figure 13). The twisted blade bit is really a twisted star bit and assists suspension of cuttings by a stirring action. These use

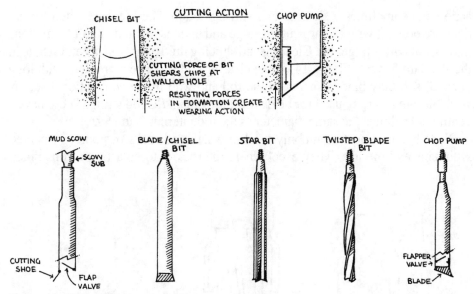

Figure 13. Cable drill bits. Reproduced from the *Australian drillers guide* by permission of the Australian Drilling Industry Training Committee.

crushing, shearing, stirring and shattering actions. The length and timing of stroke, and hence productivity, is related to ground hardness, viz.
- 57-65 strokes/min – 47.7 cm stroke, Hard;
- 50-57 strokes/min – 56 cm stroke;
- 43-50 strokes/min – 80 cm stroke;
- 35-43 strokes/min – 101 cm stroke, Soft.

3.2.2 *Tube bits*

These bits are for softer formations, and vary from just a simple tube with perhaps some hard facing to tubes with flapper valves, such as a chop pump (Figure 13) or a mud scow. They rely on shearing or displacement.

3.3 ROTARY BITS

The principal forces used are thrust, rotation, and hydraulic. To use these forces there are two basic bit types in use.

3.3.1 *Blade bits*

These are also widely known as drag bits (Figure 14). They use rotary and sometimes hydraulic forces to impart primarily a cutting, grinding, and abrading action and sometimes a sluicing and stirring action. Hard, consolidated formations are best cut with a sharp cutting edge and unconsolidated or soft formations may be cut with a finger bit. Clearing and sluicing fluid exits from a central hole in the standard format (Figure 14). Hole clearing can often be a problem with long peels of soft clay binding between the rod and wall in the conventional set up. Kick bits are a variety used for blade drilling with an *RC* rod string. These have a central hole almost the same diameter as the rod internal diameter.

Blade bits are most commonly used in weathered rocks or poorly cemented sediments and the like. They excel where the rock is too plastic for a hammer.

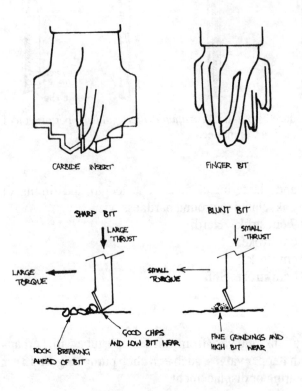

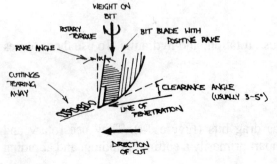

Figure 14. Blade or drag bits. Reproduced from the *Australian drillers guide* by permission of the Australian Drilling Industry Training Committee.

These bits are, however, prone to cause deviation if rods are not stabilised, and so are rarely used for water bores or precollars. The hole can certainly be held straighter with stabilising bar and reaming collars.

3.3.2 *Air coring bits*

These are very much like diamond core bits but with much larger cutting stones or inserts. They operate like an annular blade bit on an *RC* rod string. The 'core' they produce is usually little better than large chips or flakes. Air coring is a most useful technique where sticky clays will bind the rods in normal circulation with a blade bit.

3.3.3 *Roller bits*

Roller bits have a variety of design parameters for different parts of the bit, each of benefit for particular situations.

Bearing type
   1. Non-sealed roller bearing lubricated by drilling fluids. These are good in air mist or foam or non-circulating water, but wear quickly with sandy mud or water.
   2. Sealed roller bearings for mud and sandy water.
   3. Bush type bearings for high load, such as with deep water or oil holes.

Position of circulation orifices
   1. A normal rock or general purpose bit will have fluids exiting the rods through a central hole above the rollers (Figure 15).
   2. A jet roller bit has the fluid directed to the face, as these bits are for soft sluiceable material (Figure 16).

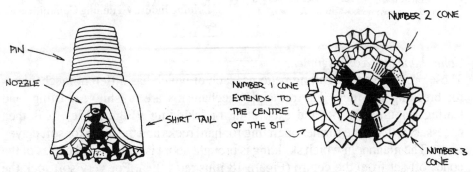

Figure 15. Tricone roller bits. Reproduced from the *Australian drillers guide* by permission of the Australian Drilling Industry Training Committee.

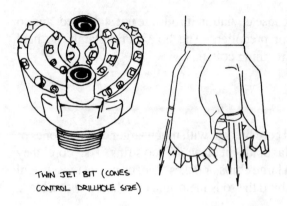

TWIN JET BIT (CONES
CONTROL DRILLHOLE SIZE)

Figure 16. Jet bits. Reproduced from the *Australian drillers guide* by permission of the Australian Drilling Industry Training Committee.

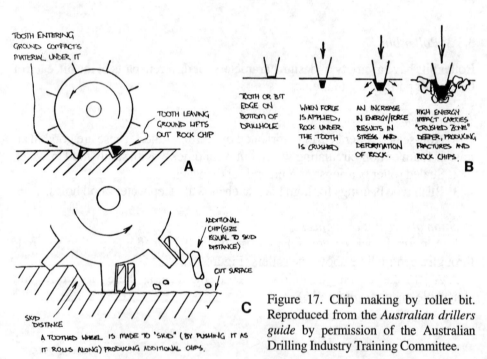

TOOTH ENTERING
GROUND COMPACTS
MATERIAL UNDER IT

TOOTH LEAVING
GROUND LIFTS
OUT ROCK CHIP

A

TOOTH OR BIT
EDGE ON
BOTTOM OF
DRILLHOLE

WHEN FORCE
IS APPLIED,
ROCK UNDER
THE TOOTH
IS CRUSHED

AN INCREASE
IN ENERGY/FORCE
RESULTS IN
STRESS AND
DEFORMATION
OF ROCK.

HIGH ENERGY
IMPACT CARRIES
"CRUSHED ZONE"
DEEPER, PRODUCING
FRACTURES AND
ROCK CHIPS.

B

ADDITIONAL
CHIP (SIZE
EQUAL TO SKID
DISTANCE)

CUT SURFACE

SKID
DISTANCE

A TOOTHED WHEEL IS MADE TO 'SKID' (BY PUSHING IT AS
IT ROLLS ALONG) PRODUCING ADDITIONAL CHIPS.

C

Figure 17. Chip making by roller bit. Reproduced from the *Australian drillers guide* by permission of the Australian Drilling Industry Training Committee.

Cone shape and axis alignment

These provide various rolling patterns, each of which has different mechanisms for breaking the rock. The principal mechanisms are crushing, twisting, and tearing (Figure 17a, b), and excavating or 'skidding' (Figure 17c). It is then necessary to have the former operating for hard rocks and the latter for soft clayey rocks. The manner in which skidding is brought about is by having the axes of the cones off-set from the centre (Figure 18 illustrates this). For very soft rock the off-set is about 3-4°, in medium rock 1-2°, and in hard rock zero. When skidding, we need more torque and therefore a bigger rig.

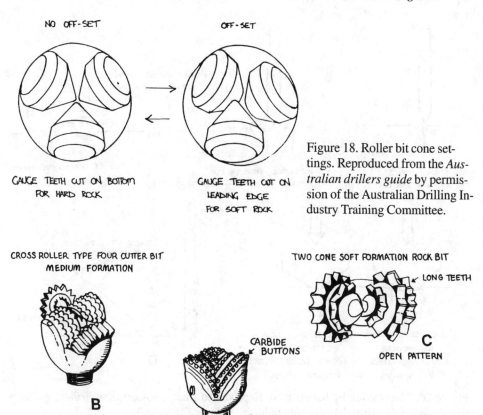

Figure 18. Roller bit cone settings. Reproduced from the *Australian drillers guide* by permission of the Australian Drilling Industry Training Committee.

Figure 19. Roller bit variations. Reproduced from the *Australian drillers guide* by permission of the Australian Drilling Industry Training Committee.

Tooth – Shape, material, height, spacing

For soft rocks, long teeth, widely spaced are best. For hard rocks small teeth, closely spaced, are best. The teeth should also have either tungsten carbide inserts or have hard facing. Hardened teeth may also be useful for softer abrasive rock (Figure 19a).

The number of rollers can also be varied according to the needs (Figure 19b, c), but a tricone is most common.

3.4 HAMMER BITS

These bits rely upon crushing, impacting, spalling and shattering as shown in Figure 20a, b. Figure 20c, d shows the effects of bit sharpness, the principles of which can be applied to other bit types also. Chip making is assisted by indexing

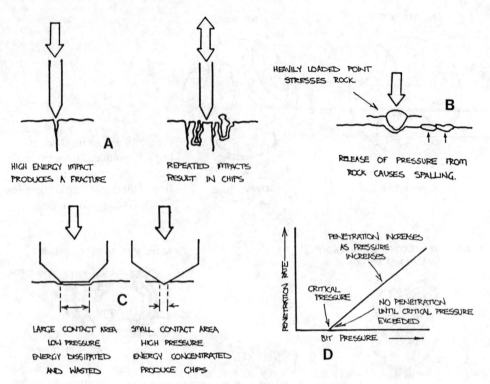

Figure 20. Chipmaking by hammer bit. Reproduced from the *Australian drillers guide* by permission of the Australian Drilling Industry Training Committee.

rotation, which is rotation that occurs between hammer blows, so that bits are not worn by being dragged on the bottom under load. Flushing clears the chips before the next impact and each impact strikes a different point.There are three major types in use:

3.4.1 *Chisel bit*

These have a single blade often as an integral part of the rod and have a tungsten carbide insert (Figure 21a). Circulation is from a single small hole behind the blade. Normally used only for holes to 35 mm diameter and mostly in jack-hammers.

3.4.2 *Cross bit*

These are also called star bits and have four chisel shaped segments (Figure 21b). The segments are almost at right angles. Circulation is through a central hole on the bit face. The most common sizes seen are from 50 mm to 150 mm. These work

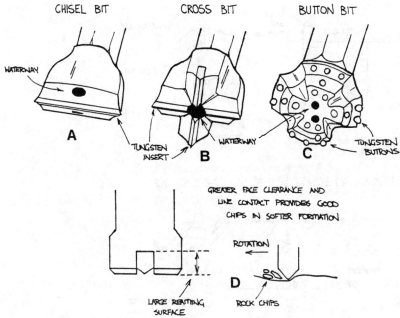

Figure 21. Hammer bit types. Reproduced from the *Australian drillers guide* by permission of the Australian Drilling Industry Training Committee.

better than button bits in softer rock where shattering is not required (Figure 21d). They work poorly in fractured rock in which the rotation becomes impeded by the blade catching in cracks.

3.4.3 *Button bit*

These are normally almost flat faced bits with a number of tungsten carbide impregnated buttons with a hemispherical shape exposed (Figure 21c). Circulation is through a number of holes at the face. The size, shape and pattern of the buttons varies according to the rock type, hammer and hole size. These are by far the most common hammer bits seen today. They range in size from 90 mm to 457 mm in diameter.

3.5 DIAMOND BITS

Diamond bits consist of a steel blank, to the lower end of which a metallic matrix is sintered, and on/or within this matrix are diamonds (natural and synthetic, or occasionally chips of tungsten carbide or silicon carbide as special alternative (Figure 22a).

A new development by De Beers is the polycrystalline diamond (*PCD*). As

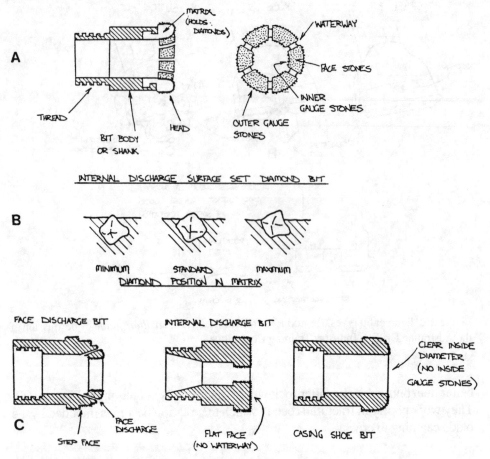

Figure 22. Diamond core bit construction. Reproduced from the *Australian drillers guide* by permission of the Australian Drilling Industry Training Committee.

reported by Clark & Shafto (1987), there are two types being developed:

Syndrill. Cobalt-Diamond fusion with tungsten carbide layers. This can be brazed to bit shells at low temperature.

Syndax3. Diamond/Silicon Carbide fusion with no layering.

Both types are manufactured at high temperature and pressure as small discs which are cut by lasers to a variety of shapes as required (e.g. cube or other pyramidal shapes as shown in Figure 23). These are set as for other surface set bits in a sintered metal matrix. As they have very precise shape and orientation they cut primarily by shearing (see Figure 24), giving much greater penetration rates than natural diamonds, particularly in softer rocks. The properties of this material are shown in Table 4.

Figure 23. Polycrystaline diamond. By permission of Longyear Australia.

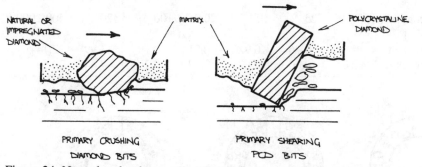

Figure 24. Natural and polycrystaline diamonds at work. Reproduced from the *Australian drillers guide* by permission of the Australian Drilling Industry Training Committee.

3.5.1 *Surface set bits*

These were the normal bits being used up to about 1980, and are still used for special purposes such as control of hole deviation. A wide range of diamond sizes and distribution is used according to the rock being drilled (see Table 5 in the section on rock drillability and also Figure 25). The diamonds are somewhat larger than within impregnated bits, and are set very accurately in the matrix so that less than 50% is exposed (Figure 22b). The quality of diamonds is also carefully graded, and range from *E* for very soft rocks through *A*, *AA*, *AAA*, to *premium* (*AAAA*) for very hard rocks.

The size of diamonds being used varies from about 2.5 mm to 1.0 mm (8 stones per carat to 120 stones per carat) Generally speaking the smaller stones are for harder rock.

The circulating fluids pass variously through the face of the bit, as can be seen

Table 4. Mechanical properties of Syndax3 and other hard materials. From Clark & Shafto (1987).

Property	Al_2O_3	Al_2O_3 + TIC	Sialon	Tungsten carbide (K10)	Syndax3	Diamond
Density (g.cm^{-3})	3.91	4.28	3.20	14.70	3.43	3.52
Compressive strength (GPa)	4.00	4.50	3.50	4.50	4.74	8.68
Fracture toughness (MPa.m$^{0.5}$)	2.33	3.31	5.0	10.8	6.89	3.4
Knoop hardness (GPa)	16	17	13	13	50	57-104
Young's modulus (GPa)	340	370	300	620	925	1141
Modulus of rigidity (GPa)	153	160	117	258	426	553
Bulk modulus (GPa)	243	232	227	375	372	442
Poisson ratio	0.24	0.22	0.28	0.22	0.086	0.070
Thermal expansion coeff. ($10^{-6}.°K^{-1}$)	8.5	7.8	3.2	5.4	3.8	1.5-4.8
Thermal conductivity (W.m$^{-1}.°K^{-1}$)	25	35	20-25	100	120	500-2000
Wear coefficient	0.76	0.92	0.91	0.79	2.99	2.14-5.49
Thermal shock resistance	0.60	0.62	3.17	10.2	7.44	24.5

3-5 STONES PER CARAT   10-14 STONES PER CARAT   80-100 STONES PER CARAT

Figure 25. Diamond distribution in bit face. Reproduced from the *Australian drillers guide* by permission of the Australian Drilling Industry Training Committee.

in Figures 26a, b, c, and 22c. Whilst centre discharge bits are easier to manufacture and give better clearing and cooling, they can wash soft core, so in soft ground face discharge bits are preferred.

The principal hole making actions are crushing, ploughing, and spalling (Figure 26d). It is normal practice to reclaim the diamonds after a bit is wornout (matrix has worn so much that diamonds are not being held in any more). The disadvantage that these bits have, is that with a shorter bit life, there is more rod tripping with more labour costs and downhole problems like caving, because rods are not in the hole to support the wall.

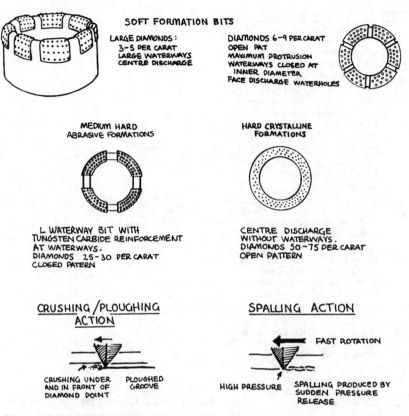

Figure 26. Diamond bit waterways and holemaking. Reproduced from the *Australian drillers guide* by permission of the Australian Drilling Industry Training Committee.

As well as the more conventional bits shown in Figure 26, there are a variety of sectional profiles with special purposes such as internal stepped, externally stepped bits, flat faced, bullnosed (Figure 27), and sometimes even noncoring bits for rotary rigs as shown in figure 28.

As a guide the following table (Table 5) shows the general characteristics of Longyear *N* size bits for various rock types. The distribution of diamonds is 20% to 30% greater on bit faces than on steps.

3.5.2 *Impregnated bits*

Here much finer diamond chips, natural and synthetic, are incorporated through-out the matrix. Again, there is a variety of diamond size and density. As an example of the effect of design variables and their relationships to rock types the following graphs (Figure 29) from Rulisek et al. (1987) are presented. 'Drilling Power' is a complex factor related to penetration rate, energy used and wear rate. The most important parameter is the hardness of the matrix. Figure 30 shows how

PROFILE	LETTER	DESCRIPTION
		Multistep: (Standardized by DCDMA). Very popular bit for wireline drilling—good penetration and stability in all but very broken formations.
	B	Semi-round: (Standardized by DCDMA). The most commonly used of non-step core bits. Exceptional strength in very hard broken ground. Requires high bit loads.
	A	Fully round: (Standardized by DCDMA). Strong setting at gauges.
	W	Part-round Profile: Very strong O.D. gauge, good for rough drilling, collaring.
	X	Narrow Pilot: Particularly good core recovery in soft, friable formations, especially when used with face discharge waterways. Good stabilization. Low vibration.
	V	O.D./I.D. Steps: Used in special applications requiring stable straight hole drilling.
	M	Tapered Pilot: Good in most formations. Stable with strong I.D. and O.D.

Figure 27. Surface-set bit crown profiles. By permission of Longyear Australia.

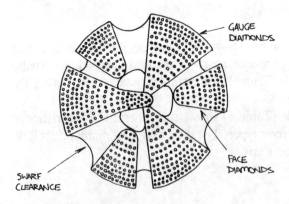

GAUGE DIAMONDS

SWARF CLEARANCE

FACE DIAMONDS

Figure 28. Diamond set rotary bit. Reproduced from the *Australian drillers guide* by permission of the Australian Drilling Industry Training Committee.

and why bits wear, and the reasons why matrix hardness must be matched to the hardness of the rock being drilled. Matrix can be pseudo-hardened by increasing the number of diamond particles.

They usually have a flat or near flat face or crown profile, and are not suited to a stepped profile because they wear the edges off too quickly. These bits are used

Table 5. Surface set diamond bits – bit design, diamond size and quality related to rock type. By permission of Longyear Australia.

Rock type	General condition	Cutting character	Diamond size*	Diamond quality	Bit profile
Soft					
Shale		Muds up	10/20	Medium or	Multistep
Chalk	Solid	Average	15/25	lower price	
Talc		Abrasive	25/40	per carat	
Gypsum					
Marble		Muds up	10/20		
	Fractured	Average	15/25		
		Abrasive	40/60		
Medium					
Siltstone		Muds up	25/30		
Limestone	Solid	Average	40/50		
Sandstone		Abrasive	60/80		
Slate					
Schist		Average	40/50		
Basalt	Fractured	Abrasive	60/80		
Monzonite Porphry					
Andesite					
Hard					
Granite		Average	25/50		
Garnet schist	Solid	Abrasive	25/50		
Gneiss					
Rhyolite		Average	25/50		
Gabbro	Fractured	Abrasive	40/80		
Diorite					
Very hard					
Granite		Average	80/150	Premium	
Quartzite	Solid	Abrasive	80/150	price per	
Chert				carat	Flat faced
Jasper		Average	80/150		
Taconite	Fractured	Abrasive	80/150		

*Stones per carat (0.2 grammes)
Note: Fine grained hard rock such as jasper and chert are best drilled with impregnated bits.

until the matrix and the diamonds are worn away, so there is no reclamation.

Impregnated bits have been classified by Longyear, one of the leading manufacturers, as *series* 1 for softest rocks up to *series* 10 for hardest rocks as can be seen in Table 6. Tables from other manufacturers follow a similar scheme, but all are constantly being modified.

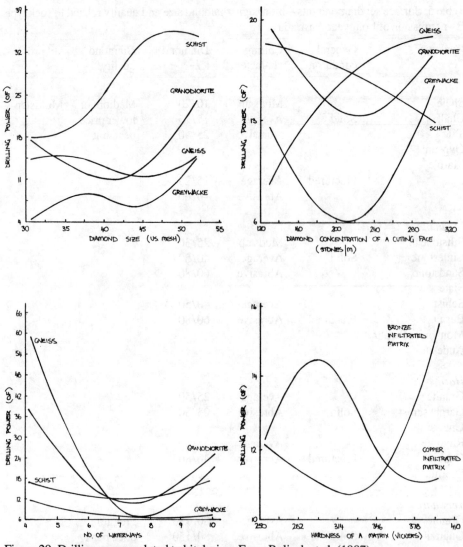

Figure 29. Drilling power related to bit design. From Rulisek et al. (1987).

As the drilling with impregnated bits makes up by far the bulk of diamond drilling, the following guidelines on drilling with impregnated bits are quoted from the Longyear Diamond Products Field Manual (2nd Edition):

'When using surface set bits, most drillers adopt the practice of choosing the rpm and bit weight they wish to use and then adjust the fine feed control to maintain this particular weight as minor variations in the formation are encountered.

As a rule, impregnated diamond bits require higher rotational speeds to achieve penetration rates comparable with those of surface set bits. This is simply because

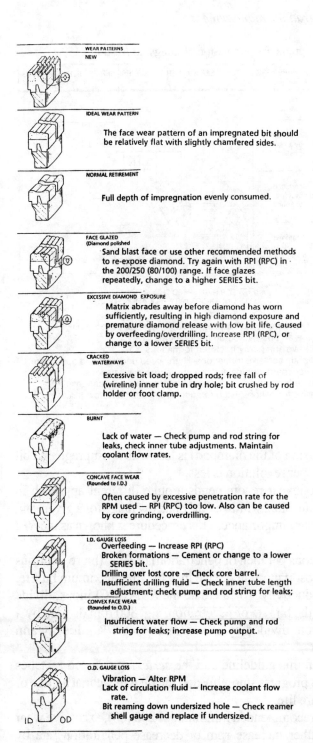

WEAR PATTERNS
NEW

IDEAL WEAR PATTERN

The face wear pattern of an impregnated bit should be relatively flat with slightly chamfered sides.

NORMAL RETIREMENT

Full depth of impregnation evenly consumed.

FACE GLAZED
(Diamond polished)

Sand blast face or use other recommended methods to re-expose diamond. Try again with RPI (RPC) in the 200/250 (80/100) range. If face glazes repeatedly, change to a higher SERIES bit.

EXCESSIVE DIAMOND EXPOSURE

Matrix abrades away before diamond has worn sufficiently, resulting in high diamond exposure and premature diamond release with low bit life. Caused by overfeeding/overdrilling. Increase RPI (RPC), or change to a lower SERIES bit.

CRACKED WATERWAYS

Excessive bit load; dropped rods; free fall of (wireline) inner tube in dry hole; bit crushed by rod holder or foot clamp.

BURNT

Lack of water — Check pump and rod string for leaks, check inner tube adjustments. Maintain coolant flow rates.

CONCAVE FACE WEAR
(Rounded to I.D.)

Often caused by excessive penetration rate for the RPM used — RPI (RPC) too low. Also can be caused by core grinding, overdrilling.

I.D. GAUGE LOSS

Overfeeding — Increase RPI (RPC)
Broken formations — Cement or change to a lower SERIES bit.
Drilling over lost core — Check core barrel.
Insufficient drilling fluid — Check inner tube length adjustment; check pump and rod string for leaks;

CONVEX FACE WEAR
(Rounded to O.D.)

Insufficient water flow — Check pump and rod string for leaks; increase pump output.

O.D. GAUGE LOSS

Vibration — Alter RPM
Lack of circulation fluid — Increase coolant flow rate.
Bit reaming down undersized hole — Check reamer shell gauge and replace if undersized.

ID   OD

Figure 30. Impregnated diamond bit wear patterns, problems and remedies. By permission of Longyear Australia.

Table 6. Impregnated bit selection guide. By permission of Longyear Australia.

ROCK TYPE	ROCK CONDITION	ROCK HARDNESS GUIDE	SERIES 2	SERIES 4	SERIES 6	SERIES 7	SERIES 8	SERIES 9	SERIES 10
ULTRA HARD Jaspilite Quartz Ironstone Chert	Competent Non Abrasive	Scratches steel							
VERY HARD Quartzite Gneiss Rhyolite	Competent Non Abrasive	Scratched by a file or steel						TK	
Granite Diorite Porphyry	Broken Abrasive						TK		
HARD-MED. HARD Andesite Basalt Gabbro Pegmatite	Competent Non Abrasive	Scratched by a file or knife							
Peridotite Schist Weathered-granite	Broken Abrasive								
SOFT-MEDIUM Sandstone Tuff Limestone Shale	Competent Non Abrasive	Scratched by a fingernail							
Calcite Gypsum Serpentinite	Broken Abrasive								

☐ MOST APPROPRIATE   ▨ ALTERNATIVE   TK FOR HIGH SPEED THIN KERF UNDERGROUND APPLICATIONS ONLY

Series 2   — For very abrasive, fractured formations. Durable, versatile, general purpose bit.
Series 4   — For medium grain, abrasive and partly fractured formations.
Series 6   — For medium to hard, abrasive and partly fractured formations.
Series 7   — For hard to very hard, slightly abrasive formations. High speed underground applications.
Series 8   — Free cutting, for hard to very hard, competent and non-abrasive formations.
Series 9   — For very hard, non-abrasive formations. High speed underground applications.
Series 10 — For ultra-hard, non abrasive formations. High rotation speed desirable, low thrust.

the diamond exposure (protrusion of the diamond) is less with an impregnated bit and consequently penetration per revolution is less.

With impregnated bits, Longyear recommends a quite different approach in which penetration rates are controlled within a fairly narrow range for a given rpm and the bit weight is of secondary importance. This procedure is known as the *RPI* (*RPC*) method of drilling.

The *RPI* index (bit revolutions per inch of penetration) or *RPC* (bit revolutions per cm of penetration) is a most important device in achieving maximum bit life, lowest bit cost and good productivity. To calculate the *RPI* (*RPC*) index, divide the rotational speed (rpm) of the bit rate of penetration, e.g. 800 rpm divided by 4 in/min = 200 *RPI*, or (800 rpm divided by 10 cm/min = 80 *RPC*). Ideally aim for 200-250 *RPI* (80-100 *RPC*).

Providing you work within this guideline and the *series* bit selected matches the formation, drilling should progress smoothly and the bit will wear at a more or less constant rate over its entire life.

If *RPI* (*RPC*) is below the recommended minimum of 200 (80), excessive wear will occur so you should either increase rpm or decrease penetration rate by reducing bit weight. If ground conditions or drill limitations prevent you from making these adjustments, you should change to the bit next down the *series*, e.g. from *series* 10 to *series* 8.

If the *RPI* (*RPC*) index is much above the recommended maximum of 250 (100), the bit may 'polish' so you should reduce rpm or increase penetration rate by increasing bit weight. If rpm or weight cannot be altered, you should change to the next bit up the *series*, e.g. from *series* 6 to *series* 8. 'Polishing', 'glazing' or 'closing up' are terms commonly used to describe a condition where the face of the bit becomes metal bound and no diamond points protrude from the matrix to cut the rock. Penetration virtually ceases and it becomes necessary to 'strip' the bit down the hole or otherwise re-expose the diamond. It is of the utmost importance, in order to avoid polishing, that the driller should *keep the bit cutting.*

Although we have stated earlier that bit weight is of secondary importance when drilling with impregnated bits, we realize that it can become an important factor in some circumstances, especially where the limitation of the down-hole tools to withstand high thrusts is being approached, or where deviation control is of prime importance.

In such cases it is suggested that a higher than normally recommended *series* bit should be used, exercising some discretion concerning penetration rates. This will tend to diminish the deviation or down-hole equipment problems at some expense in higher bit costs.

The bit weights shown ......(are) considered normal for a given size of core barrel. If you exceed the maximum weight listed, you can expect deviation to arise, excessive wear of core barrels and rods, and even down-hole failures.

Regulation of fluid pump output can be a useful technique for assisting the drilling of very hard siliceous formations. Under these conditions, pump outputs should be reduced ...... and this will assist in a minor buildup of cuttings at the bit which will, in turn, abrade the matrix.

Customers are encouraged to use cutting oils or lubricating fluids with *series* impregnated bits'.

Figure 30 shows impregnated bit wear patterns and what they mean.

3.5.3 *Soft formation bits*

As one would expect, in soft plastic rocks such as clays, the small 'teeth' which normal diamonds exhibit cannot cut or tear the formation effectively. Larger teeth

Figure 31. Soft formation coring bits. Reproduced from the *Australian drillers guide* by permission of the Australian Drilling Industry Training Committee.

or blades are needed with the cutting edge being tungsten or silicon carbide, or polycrystalline diamond.

This can be surface set or used as blade tips in annular blade bits (Figure 31). Annular roller bits are another alternative. All these bit varieties can be used to get core with rotary drills, as high rotational speed is unnecessary in soft rocks.

CHAPTER 4

Rock drillability and stability

The drillability of rock depends not only on the physical properties of the rock, but also the type of drill being used and the drilling parameters such as rotation speed and feed rate etc. Drillability is not just penetration rate or dollars per metre, it is also how satisfactory is the product (e.g. how straight and clean is the waterbore?, or what was the core recovery in a diamond hole?).

Firstly we shall examine the physical properties of rock which have some effect on drillability.

Crushing strength
Defined as the pressure a rock sustains before breaking and this is related to:
  – Grain hardness and strength;
  – Grain bond strength;
  – Porosity;
  – Weakness planes.

Toughness
This is a measure of how difficult it is to pull a rock apart (i.e. how resilient), and is related to:
  – Grain shape;
  – Grain bond;
  – Fissility; Tenacity.

Chip separation
This is how readily the cuttings are cleared from the face, and is related to:
  – Pore pressure;
  – Permeability.

Abrasiveness
This is the ability to wear downhole tools, and is related to:
  – Grain hardness;
  – Grain shape.

Table 7. Drillability properties of some common rocks.

High strength	Plastic	Fissile	Very abrasive
Quartzite	Clay	Schist	Quartzite
Rhyolite		Shale	Rhy. tuff
Granite			Granite
Shale	Dolerite		
Schist	Granite		Basalt
Chalk	Chert	Granite	Mudstone
Low strength	Brittle	Non-fissile	Not abrasive

Table 8. Drill usage related to rock type and drillability.

Grain size	Rock types	Drill types
High crushing strength & abrasive		
Coarse	Granite, quartz porphyry	Pc (heavy), D
	Pegmatite, gneiss	Pu (50-100 mm diam.)
Medium	Quartzite, quartz schist	Pd (100-150 mm diam.)
Fine	Obsidian, perlite, rhyolite, jasper, chert	
High crushing strength & less abrasive		
Coarse	Diotite, gabbro, syenite	Pd, D, Rr (large)
Medium	Andesite, dolerite, quartz-mica & chlorite schists,	Pc (small holes)
	sandstone, conglomerate, fine basalt, trachyte	
Moderate crushing strength & abrasive		
Variable	Grits, sandstone, siltstone, agglomerate, tuff, ash, marl	Rr (medium), Pd, D
Moderate crushing strength & non-abrasive		
Variable	Limestone, mudstone, shale, phyllite, slate	Rr (light), Pd, D
Low crushing strength & abrasive		
Variable	Siltstone, argillaceous and	Rr & b (light), D
	Calcareous sandstones,	Pc, Pu, Pd (light)
	Non-indurated sediments	A
Low crushing strength & non-abrasive		
Variable	Mudstone, shale, marl, coal	Rb, D, A
	Chalk, decomposed rock, coral	Pc (if boulders)

Legend for drill types: A = Auger, P = Percussion, c = Cable, u =Up hole, d = Down hole,
R = Rotary, r = Roller, b = Blade, D = diamond.
Note: Rock types are indicative only – listing is not meant to be exhaustive.

Table 7 shows how some common rock types may rate for these properties.

Table 5, from data supplied by Longyear (Aust) Pty. Ltd, relates surface set diamond bit types to rock types. Modern impregnated bit usage is shown in Table 6 from Longyear (Aust) Pty. Ltd.

Table 8 relates diamond, rotary and percussion drilling to the drillability of rocktypes. Although it is not exhaustive, nor does it allow for the sometimes great range of drillability of some rocks, this is a very useful guide.

Firstly, it is important, in reading these tables, to keep in mind that a property which makes one drilling method easy, may well make it difficult for another. For example, meta-dolerite, with its interlocking felt textured amphiboles, is difficult to hammer (adsorbs shock) but is easy to core. A glassy brittle rhyolite on the other hand will be the opposite. A soft puggy clay will probably cut well with rotary blades but cannot be hammered.

Secondly, we should consider stability, which is the extent to which hole walls require support by mud or casing. The properties which relate to hole stability are: Consolidation; Fractures/joints etc.; Weathering; Formation sensitivity (to pore pressure, contact or invasion by various fluids).

These properties are considered in Chapters 5 and 6.

CHAPTER 5

Circulating fluids and additives

The principal purpose of a circulating fluid is to remove the cuttings from the face of the bit and to provide cooling of the bit. The quantity of fluids required will depend on the size of hole, rod to wall clearance, and type of bit being used. For diamond drilling it will range for example, from 8 litres per minute for *AQ*, to 70 litres per minute for *PQ* surface set bits, and for impregnated bits it will be 10% to 20% more for either size.

Additives to the fluid can be of critical assistance to the successful drilling and sample recovery. These additives can account for a very significant part of the drilling costs as in most current contracts these are cost plus. Because of this, the circulating fluid and additives are now of prime concern to site geologists. The cost plus costing tends to encourage drillers to use additives to excess and indiscriminately. For example I have seen soluble oil being used at up to 10% (2%-normal, 5%-max.) and have seen mud mixtures which just incorporate everything on site, on the basis that if a little helps, a lot will be better.

The circulating fluids have the following functions:

Removal of cuttings
The flow properties must be adequate to allow transport to the surface of cuttings and any caving material. This gives the bit a clean face on which to work and provide the necessary cooling.

Cooling of downhole tools
The heat of cutting and rod string friction has to be dissipated or there will be very severe wear and deformation (melting).

Suspension of cuttings
Whilst the circulation is stopped to trip rods or pull core etc., the cuttings should stay in suspension. The gel strength must be sufficient to prevent settling.

Control formation pressure
The hydrostatic head in the hole should be high enough to prevent ingress of

formation fluids, which may cause caving or maybe a blow-out. This must be balanced by the fact that high hydrostatic pressure can cause excess loss of drilling fluids to the formation.

Lubrication
The slippery mud and/or soluble oil will decrease friction between rods and casing or wall and lubricate roller bit bearings.

Wall support
There are three ways that circulating fluids do this.
   - *Plastering*. Mud will gradually form a filtercake on the wall of permeable formations which will reduce the further loss of fluid. However if the cake builds up too thick it will cause rod jamming and blockage of circulation.
   - *Hydrostatic pressure*. If larger than the formation pressure, it will help avoid spalling from the walls.
   - *Chemically*. Reaction with wall rocks, such as salt to prevent further wetting.
   The principal drilling fluids used are air and water. Hammer drills use an air circulation system, diamond rigs use a water circulation system and rotary rigs use both. There are of course some special techniques which differ from these basic uses.

5.1 AIR BASED CIRCULATING SYSTEM

To the air may be added water, detergent, lubricating oil, and to the water may be added mud, polymers, oil, chemicals and detergent. The number of additives used is normally much more restricted than with a water based system, and the most common are:
   Water – mist – to clear hole of damp, sticky cuttings.
   Foam – helps clean hole if air velocity is not adequate, there is some rod sticking and moderate inflows. If very stiff foam is used, very little air can get out.
   Gel-Foam – Reduces volume of lost air to porous and fractured formations.
   Removes large amounts of formation water.

5.2 WATER BASED CIRCULATING SYSTEM

This is a three phase system.
   1. Carrier – water ± oil emulsion;
   2. Colloid – mud, polymer;
   3. Inert – cuttings, barite, etc.

The major part of mud control is the colloid, which can be improved by the addition of chemicals or damaged by contaminants. Natural clays encountered in a water-only (no additives) hole, may provide ideal mud properties for that situation. Differing rocks, depths, water and tools, all mean different combinations of required additives, and of course air may be an additive to a water system.

The following is a list of common additives:

1. Soluble oil – 0.5%-5.0% concentration will improve penetration rate and bit wear in hard rocks (some alcohols and chemicals may have a similar effect).

2. Clay muds – natural muds such as Bentonite, to increase the viscosity, form filter cake to stabilise walls and reduce lost circulation.

3. Polymers – increase viscosity, decrease filtration, decrease cuttings dispersion, flocculate drill cuttings. They will commence decomposition after 72 hours unless preservative is added or pH of 10 is maintained.

4. Soft soap – reduction of friction.

5. Dispersant – thinners make mud less viscous.

6. Colloid – increases gel strength and lubrication.

7. Barite – increases specific gravity of mud to increase downhole pressure to restrict flows of water, etc.

8. Flocculant – precipitates suspended solids. Could use gypsum, lime or cement, as well as brand products.

9. Sealant – prevents circulation losses. Usually grains, flakes and/or fibres which will be carried by the fluid and shortly swell to hold in cavities, etc.

10. Diesel – clears blockages, excess grease, etc.

11. Salt – seals clays in hole wall. Potassium Murate (KCl) is most common.

12. Anticorosive – used with saline fluids.

13. Foaming Agent – improves hole cleaning.

14. Defoamer – increases surface tension.

15. Wetting Agent – decreases surface tension, retards cutting breakup, promotes surface settling of cuttings, reduces rod sticking.

When water inflows are encountered in holes with air circulation, usually much water is blown out of the hole collar both by the differential pressure, and by the airlift principle of expanding air bubbles as they near the surface under a lower pressure regimen. When water flows are intersected in diamond drillholes with a water based circulation then the water is blown, not out of the collar as normally, but back into the formation, as usually the hydrostatic head is greater than the formation pressure. This means that circulation losses may signal the presence of a significant aquifer, and should be noted.

5.3 CEMENTING

Various types of cement and alternatives may be pumped down the hole for three principal reasons.

- To restrict water flows into or out of a hole;
- To hold together caving ground;
- To provide a stable base from which to wedge.

It should be emphasised at this point that cementing should be a last resort. It can be very time consuming, particularly if the local drilling or ground water acts as a retardant.

The grouting of underground holes to prevent flows, or engineering holes at dam sites and the like, to block potential leaks and to improve the building foundation stability, are specialised activities and are not elaborated upon.

The most common problem requiring cement or something similar in normal exploration holes is loss of water return. This can usually be best rectified by additives other than cement. These additives usually are products which will swell in water. As well as brand name chemicals there are a number of natural products available such as rice, horse manure, sawdust, oatmeal, and cotton seed hulls. More common brand name products are Kwik Seal, Hyseal, and Micatex.

To hold caving ground, the three products which are normally used are Cement Fondu (hard and quick), Gypseal (quick but soft), and Portland Cement (with accelerators such as calcium chloride or even retardants; has variable speed of setting and moderate strength).

Where it is desired to cement other than at the bottom of the hole it is common for a plug (piece of wood or bag or specially made plug) to be inserted and to cement upwards from this. As cement will 'float' on water due to the differential friction on the wall between water and cement, the hole may have no plug at all, although it will take a little more cement, as some cement will be 'immersed'.

CHAPTER 6

Downhole drilling problems

The aim of this chapter is to introduce the non-driller to the main problems related to the characteristics of the rocks being drilled and the drilling fluids being used. Where relevant, the usual methods of solving these problems will be introduced. Being aware of these problems will be of considerable help in hole planning and management. If some of these problems can be anticipated, then holes can be relocated, or drillers warned, to enable corrective or preventative action to be taken. In addition to the problems associated with rock type variations, there are a number of unusual features or cultural hazards which one should note, as shown in Figure 32.

6.1 DIFFERENTIAL PRESSURE STICKING

This causes rods to be jammed. The following diagram (Figure 33) illustrates the principles involved. The conditions necessary are:
  – Hole hydrostatic pressure exceeds formation pore space pressure;
  – Formation is permeable;
  – Thick, poor quality filtercake, is built up by continuing fluid loss to the formation;
  – Rods are not kept moving.
Rods are normally released by reducing pressure by introducing low density and low viscosity fluids or air (e.g. clean water or diesel are possible). Rods could also be jarred.

6.2 KEY-SEAT STICKING

This is thought to be one of the major causes of stuck rods in deep holes. Key seats appear on the upper side (or occasionally on the bottom side) of a hole at sharp bends, where rods wear a groove into the wall which increases friction and can catch the bit on return. Figure 34 illustrates the problem, and the tools used to solve it.

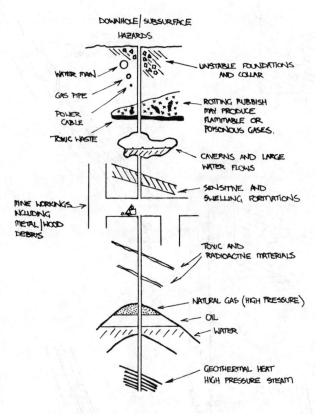

Figure 32. Downhole Hazards. Reproduced from the *Australian drillers guide* by permission of the Australian Drilling Industry Training Committee.

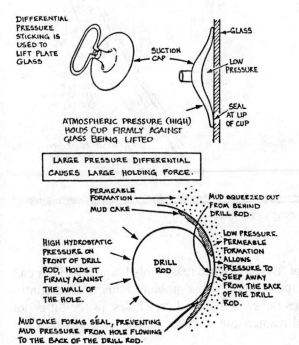

Figure 33. Differential pressure sticking. Reproduced from the *Australian drillers guide* by permission of the Australian Drilling Industry Training Committee.

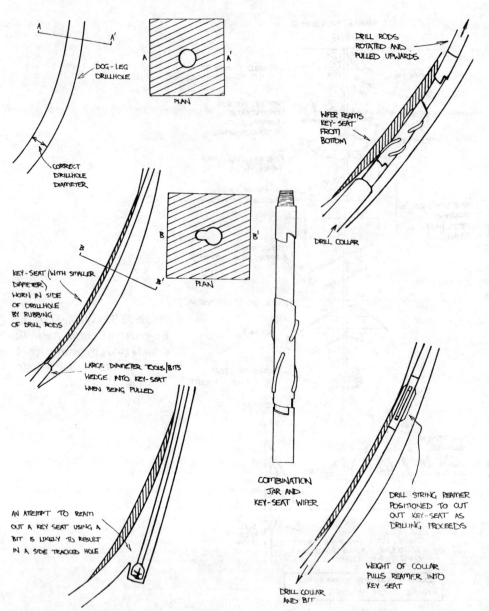

Figure 34. Key seat sticking. Reproduced from the *Australian drillers guide* by permission of the Australian Drilling Industry Training Committee.

Key-seats can be recognised if rods are tripped reasonably frequently. They can be removed by reaming with special tools. This problem is more common in rotary and percussion drill holes rather in diamond core holes, as the bit is not much larger than the rods, and reamers are included as part of a standard rod string.

6.3 FORMATION CHARACTERISTICS

6.3.1 *Problem shales*

Some fine grained sediments are fluid (water) sensitive. This depends on their particular clay mineralogy. There are three symptoms that can be readily observed, which are: swelling, 'dissolving' and spalling (due to high pore pressure). These can all be avoided to some extent by casing the hole as drilled, or drilling with air only, or most commonly by diamond drilling, using saline muds which insert preferably potassium or sodium atoms into the clay molecular lattice at the hole wall interface, which prevents water going further into the country rock.

6.3.2 *Boulder formations*

Coarse gravels, which are unconsolidated or non-cohesive, means that the rock being drilled cannot be held fast whilst being cut. Cable drills are reasonably effective, or possibly high energy downhole hammers; often explosives are used to crack the boulders a little. Casing invariably must follow closely, or even precede the bit.

6.3.3 *Permeable formations*

Permeability will cause variable fluid loss according to:
1. Absolute permeability.
2. Hydrostatic Pressure.
3. Size of openings (i.e. pore space, cracks or caverns).
4. Low strength formations in which hydraulic pressure opens up cracks, usually due to holes being over-pressurised, then becoming permeable.

The popular solutions to these problems with circulation are:
– Wait for mud to gel in formation pores;
– Drill on with no fluid return;
– Add special solids to block pores and cracks;
– Gunk squeeze, cement, or case;
– Use low density fluids (air or foam);
– Change to cable drill tools and casing.

6.3.4 *Fractured or blocky formations*

These cause hole caving, sometimes through back pressure (hydrostatic pressure in the hole is less than formation pressure), and sometimes from lack of support. Joints, fractures etc. at right angles to the core will cause far less trouble than acute angles which allow broken core to ride over and wedge. Such ground will also cave easier. Smaller diameter and near vertical holes cause less problems, and

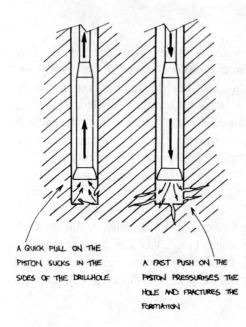

THE BOTTOM HOLE ASSEMBLY IS
LIKE A PISTON...

A QUICK PULL ON THE
PISTON SUCKS IN THE
SIDES OF THE DRILLHOLE

A FAST PUSH ON THE
PISTON PRESSURISES THE
HOLE AND FRACTURES THE
FORMATION

Figure 35. Caving by pressure inequilibrium. Reproduced from the *Australian drillers guide* by permission of the Australian Drilling Industry Training Committee.

diamond drill holes are better, as there is less rod clearance and a flush jointed rod string.

Caving can cause hole deflection, as well as the obvious problems of jammed rods, and bit wear when drilling caved rubble after rods are tripped etc. The principal solutions are high gel strength muds to stabilise the ground, and with careful rod and core pulls, so that pressure equilibrium is not disturbed (Figure 35). If serious and localised caving occurs, that part of the hole may be cemented and redrilled.

6.3.5 *Pressurised formations*

These situations are mostly encountered in oil, gas or water holes, although they are not uncommon in underground drilling. It is important to have the necessary blow-out preventers and plugs available.

6.3.6 *Poor coring formations*

The more common problems are due to the following:
1. Very hard abrasive rock – low bit life.
2. Very tough rock – low hammer penetration rate.
3. Fractures or swelling – core jams in tube.
4. Low strength, friable – fluids erode core.

6.3.7 *Poorly consolidated ground*

When drilling through loose overburden it is often convenient to run casing at the same time as the hole is being made (casing advances with the bit). There have been a number of ways of doing this in use over the years.

One of the oldest and still used in the Malaysian tin scene is the people powered Banka Drill. Here the casing is clamped to a platform on which a crew of two or four walk around whilst turning the rods. This continued force pushes the casing down as the hole is made.

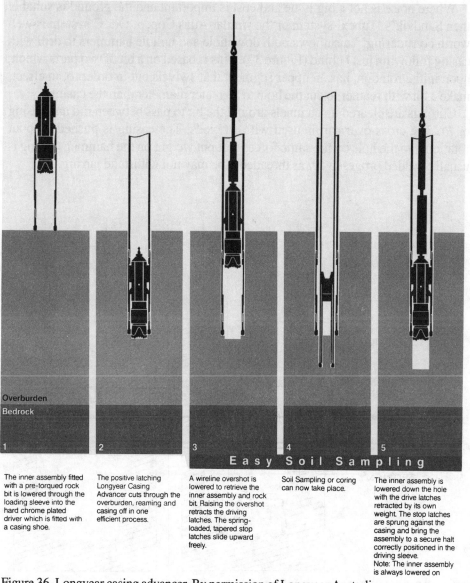

Overburden

Bedrock

1 2 3 4 5

E a s y   S o i l   S a m p l i n g

The inner assembly fitted with a pre-torqued rock bit is lowered through the loading sleeve into the hard chrome plated driver which is fitted with a casing shoe.

The positive latching Longyear Casing Advancer cuts through the overburden, reaming and casing off in one efficient process.

A wireline overshot is lowered to retrieve the inner assembly and rock bit. Raising the overshot retracts the driving latches. The spring-loaded, tapered stop latches slide upward freely.

Soil Sampling or coring can now take place.

The inner assembly is lowered down the hole with the drive latches retracted by its own weight. The stop latches are sprung against the casing and bring the assembly to a secure halt correctly positioned in the driving sleeve.
Note: The inner assembly is always lowered on

Figure 36. Longyear casing advancer. By permission of Longyear Australia.

The latest and more realistic method for normal situations comes from Longyear. They market a casing advancing system (Figure 36), which will allow quick bit changes to enable intermittent core sampling. This allows a rotary bit to be lowered or raised by wireline inside the casing and lock into the casing at the bottom of the hole so that the bit is in advance of the casing. When locked into the casing the bit will rotate as the casing rotates.

This is an alternative to normal air circulating *RC* hammer or blade in built up areas where noise is a problem, and a standard diamond rig is therefore more desirable than say, a multipurpose rig.

Where noise is not a big issue and cost is important and the ground is suitable then Sandvik's 'Tubex' system or the similar Atlas Copco 'Odex' system is well worth considering. This allows both downhole and uphole hammers to drill with casing following just behind (Figure 37). This is based on a bit of two parts which, upon initial rotation, has an upper segment that swivels out in order to, in effect, make a bit with reamer to cut the hole to a greater diameter than the casing.

Cuttings are cleared up channels around the bit to pass between rod and casing or *RC* if a cross-over sub is used with *RC* rods. The casing is pushed without rotation into the hole by the same forces that put weight on the hammer. Casing is usually welded progessively as threaded pipe may not withstand jarring.

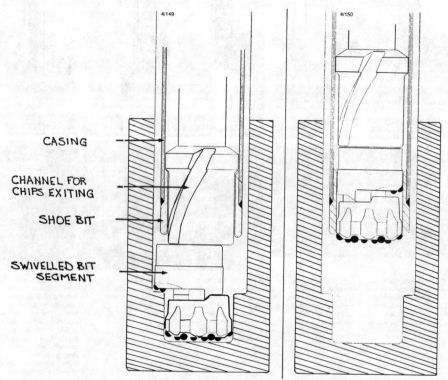

CASING

CHANNEL FOR
CHIPS EXITING

SHOE BIT

SWIVELLED BIT
SEGMENT

Figure 37. Sandvik Tubex System. By permission of Sandvik Australia.

6.4 EQUIPMENT LOSSES DOWNHOLE

On almost all drilling programmes there will be some equipment lost downhole which will require retrieval or bypass to enable the hole to continue. Sometimes there will be much lost time and this, and the equipment value, have to be evaluated to decide who is paying and how much and then what is the best method of 'fishing' in the circumstances. Whilst the drillers will be well versed in most 'fishing' techniques, the geologist can often make a contribution as the situation may have unique facets, and the more 'thinking power', the better.

The type of gear which may have to be fished are:
– Bits and bit parts;
– Pipe, rod, or casing;
– Core barrels or drill collars;
– Pump, pump components or column;
– Wire line, drilling or logging cable;
– Hand tools or junk;
– Wedging gear;
– Fishing tools.

The most common reasons for the above items to be fished are:
– Bit failure with parts coming loose such as inserts, cones, teeth or sheared bit shank;
– Joint failure;
– Pipe or rod failure;
– Hand tools, parts or blots etc. falling into hole;
– Casing failure or unscrewing;
– Failure of handling or hoisting equipment;
– Failure of Wireline;
– Stuck tools deliberately disconnected.

There are numerous ways of fishing (attaching to and removing) for the object, and each situation will suggest another variation, often dependent on what tools are available.

The more usual ways of attaching to the object are:
– Friction grip;
– Hook or set of hooks;
– Weighted or spring loaded latches;
– Taps and die collars;
– Rolling or tapered seat slips;
– Helical grapple;
– Enclosing barrel or junk basket;
– Magnet.

There are of course three alternatives to the fishing and removal of the object which are:
– Drill through;
– Drill(wedge) around;
– Abandon hole.

CHAPTER 7

Drillhole deviation

The phenomenon of drillhole deviation is well known, but some of the problems encountered are generally due to a lack of understanding of the causes and effects. This lack of understanding is evidenced by reports containing the results of poorly planned holes, e.g. acute hole to ore structure angle, consistent azimuth variation from hole to hole, and unsurveyed holes greater than 50 m deep.

Almost all drillholes deviate. A professional driller, with the aid of progress hole plots, and good knowledge of his tools and their effect on various rocks, should be able to drill very near target. He will be greatly assisted by the geologist's accurate predictions as to the rock types, their hardness and anisotrophic parameters and how these will affect the drill path. He will be able to achieve such accuracy without recourse, in most cases, to costly deflection techniques such as wedges and the like.

The geologist, on the other hand, should be able to design a hole which will enable the driller to achieve the desired result as quickly and economically as possible.

As most of the application is associated with diamond drilling or large diameter rotary, this chapter will concentrate mainly on diamond drilling. It is estimated, for example, that approximately 3 million m of exploration and mine development drilling is done each year in Australia, and if better hole planning and control this can be reduced by 10%, which I am sure is possible, then several million dollars can be saved directly and indirectly each year. There is more application to diamond drilling because the holes are deeper and a greater target is accuracy required. Much the same cause and effect is seen in rotary and hammer drilling; the main difference with diamond drilling is the number of remedial tools. There can, however, be significant savings using downhole hammer precollars and the like, if it can be done accurately.

Following some qualitative work in the decade ending in 1960, there has been, at least to the mid-eighties, an increasing amount of experimental and investigative analysis done to predict the pattern of deviation. As yet no perfect universal formulas have evolved and the rules we do have are largely empirical, and specific of each mining area.

Bluck (1978) researched deviation at three major mining camps, Mount Isa, Cobar and Mount Morgan under sponsorship of Australian Mining Industry Research Associates Limited. He developed a mathematical model of the relationship of deviation to the anisotropic strength index. He concluded that:

'For drilling in slates and shales the gradient of component lines in plots of Log_n Radius of Curvature versus $\log_n$ Angle of Incidence (of planar weakness to drillhole axis) are approximately equivalent for both *NQ* and *BQ* drilling. Accordingly relationships can be generalised.

For angles of incidence less than 40°

$$\text{Log}_n\text{Radius} = -1.0\,\text{Log}_n\,\text{Angle} + (\text{Log}_n\,\text{Radius (min)} + 3.6).$$

For angles of incidence greater than 40°

$$\text{Log}_n\text{Radius} = 1.5\,\text{Log}_n\,\text{Angle} + (\text{Log}_n\,\text{Radius (min)} - 5.53)$$

where Radius = radius of curvature of the drillhole path corresponding to a particular angle of incidence; Angle = angle of incidence between planar penetrative structure and the drillhole axis; Radius (minimum) = radius of curvature corresponding to an angle of incidence of 40 ° – i.e. maximum hole deviation'.

Bluck produced a graph relating the maximum deviation – Log_n Radius (min) to the anisotropic strength index for *BQ* drilling, and by using this the bracketed expressions in both equations reduce to constants (for the same rock and core size in any hole). This graph is reproduced as Figure 38.

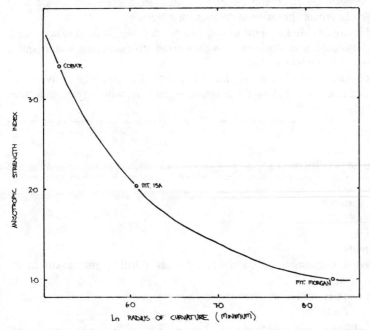

Figure 38. Anisitropic strength related to deviation. From Bluck (1978).

He also provided a set of percentage correction factors to adjust the value of the radius of curvature if other than normal *BQ* drilling. See Table 9.

The anisotropic strength index (*ASI*) was defined as the ratio of axial point load strength to the diametrical point load strength. Both normalised to strength per unit area of failure surface. This can be done on core or relatively fresh rock from mine dumps. For quick reference an estimate can be made by comparison with the three 'standards' described below in Table 10.

I suggest that whenever one becomes involved in any type of drilling, in addition to making oneself aware of new developments in this field, such as tools or techniques, drilling statistics be collected for each project area to enable more

Table 9. Correction factors for path radius of curvature. From Bluck (1978).

Drilling assembly/conditions	Variation in Log_n radius (minimum)
NQ, standard	+6%
Controlled drilling, minimum hole clearance, flat faced bit	+7%
Induced deviation drilling, new step bits, high bit pressures	–4%

Table 10. Anisotropic strength index standards. From Bluck (1978).

A.S.I.	Rock type
1.0	e.g. Mt. Morgan Quartz Porphyry. Fine to medium grained, massive, faintly banded. No discernable lineation of component materials.
2.0	e.g. Urquhart Shale (Mt. Isa). Fine to very fine grained, well bedded to laminated. The material should generally break along the penetrative structure, with frequent irregular fractures across the plane.
3.0	e.g. Great Cobar Slate. Fine to very fine grained, well to very well cleaved. The material invariably breaks along the penetrative structure, with very rare fractures across the plane.

Table 11. Drilling parameter differences.

Diamond holes have:
– Less rod to wall clearance.
– Greater bit pressure.
– Different bit design.
– Greater rod flexibility.
– Greater rock strength anisotrophy in normal target areas. (Drilling structurally more complex areas.)
– Angle of incidence to penetrative structure is less.
– Inclined holes more common.
– Higher rotational speed.

accurate prediction of drillhole deviation, under a wide variety of conditions.

There has been considerable research done on the large diameter holes used for oil drilling and whilst this may be applicable to the smaller percussion and rotary holes used in the mineral exploration industry, this is of limited use for diamond drilling, as Table 11 highlights.

Although it is difficult to assign an order of importance, there are a number of features which cause a hole to deviate both in azimuth and declination. The more important of these are:

1. Hardness of rocks – rate of penetration. Although I doubt this alone can cause deviation it does allow direct causative parameters to operate at greater rates for softer rocks.

2. Rock strength anisotropy – which is exhibited by rocks with planar textured features such as foliation and bedding. Holes tend to deviate to penetrate these features at right angles, unless the angle of incidence is very low, i.e. about 5°, in which case they may follow the feature. It has also been shown in some individual areas that rate of deviation is a function of the angle between the penetrative structure and the drill axis. Bluck (1978) found for example, that at Mt. Isa, Mt. Morgan and Cobar, the maximum deflection occurred at an angle of 40° to the foliation or bedding.

3. Anisotropic Strength Index – from work by Bluck, 1978, it has been shown that the absolute magnitude of deviation is related not only to rock strength, but to the relative strengths in different directions. Holes in well foliated schist deviate at a much greater rate than through a normal shale, which will be greater than granite. The reason for greater deviation in strongly anisotropic rocks is not absolutely clear, but is probably due to the relative cutting action on either side of the bit, and so 'wedging' occurs.

4. Active length of drill rods – by active length we mean the distance from the bit to the point of contact of the rods with the hole wall. In general the shorter this is, the greater the rate of deviation. As the principal force is largely gravity in declined holes, the hole almost invariably deflects upwards, with normal bits, thrust, etc. In hammer holes, where the hole clearance is greater, inadequate thrust can see holes 'drooping', mainly in softer formations (Figure 39).

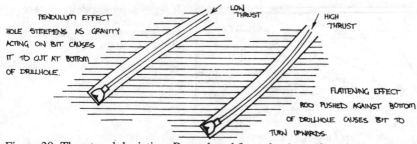

Figure 39. Thrust and deviation. Reproduced from the *Australian drillers guide* by permission of the Australian Drilling Industry Training Committee.

5. Barrel length – greater deviation occurs with shorter core barrels because this increases flexibility, and therefore decreases active length.

6. Hole size – greater deviation occurs in smaller holes – probably a function of greater flexibility of rod string, and hence smaller active length.

7. Bit type – external step surface set bits give most deviation; with flat faced and internal step surface set bits giving least, but the effects are only really seen in anisotropic rocks. With impregnated bits, a lower series bit will require more thrust and hence more deviation.

8. Direction of rotation – with normal clockwise drill rotation the azimuth of declined holes will change from left to right as the hole deepens, if the rocks are almost isotropic. Usually the steeper the hole the greater the rate of change. Often there is a spiral path of the hole, particularly in rotary and hammer holes (Figure 40).

9. Wedges – on rare occasions natural wedges can occur such as a quartz vein within a soft rock which is at a small angle of incidence.

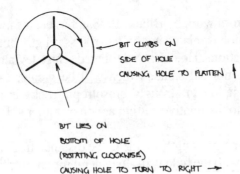

Figure 40. Rotational deviation. Reproduced from the *Australian drillers guide* by permission of the Australian Drilling Industry Training Committee.

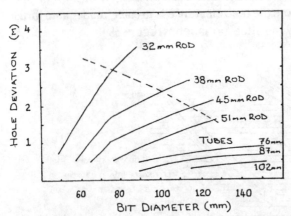

Figure 41. Uphole hammer deviation. From Schunnesson (1987).

Table 12. Summary of deviation factors.

				Minimum deviation				
Rock type	Rod diam.	wear	Core barrel length	type/wear	Bit type	Reamer position	Angle to bedding or foliation	Feed rate
Granite	HQ	New	6.0 m NQ 3.0 m NQ 1.5 m HQ	Chrome new Chrome new	Internal step Impregnated	Back end	90ø to hole	Slow
Sediments	NQ			Standard new		both		
Schists	BQ	Worn	1.5 m BQ	Standard worn	5 step external 7 step external	Front end	Acute to hole	Quick
				Maximum deviation				

Deviation of uphole hammer drillholes can been quite marked. The extra deviation is due mainly to the difference between bit and rod or coupling diameter and the non-rigidity of the rod string. As an example of this, see Figure 41 from Schunnesson (1987).

CHAPTER 8

Drillhole deflection

Firstly it is now appropriate to distinguish between deviation and deflection. Both describe the way in which a hole path varies from straight. Deviation is how it changes path naturally due to the rocks encountered, and how they interact with the composition of the rod string, i.e. it is more or less a continuous action. Deflection on the other hand is when we deliberately change this natural deviation by inserting some mechanical device, or changing the rod string. As will be seen later, there is little difference between deviation and 'natural' deflection.

As will be elaborated upon in the chapter on 'hole control', holes should, in general, be planned to use deviation to arrive at target. But having said this, there are a number of situations in which the use of wedges is the only practical solution. Some of these situations are:

   – To create daughter holes, to enable several intersections from the same collar, as a cost saving;

   – To enhance or depress natural deviation to ensure the target is intersected;

   – To bypass difficult drilling conditions, abandoned drilling equipment, or other foreign objects;

   – To obtain second intersection for improved recovery, metallurgical sample or such;

   – To force the hole path to those otherwise inaccessible locations.

8.1 ARTIFICIAL WEDGES

An artificial wedge is a device introduced to the hole, the only purpose of which is to deflect the path of the drill hole. There are basically three types in normal use with diamond drill holes in Australia:

   – Hall-Rowe;

   – Casing;

   – Clappison.

Equivalent similar types will be found with different names elsewhere in the world.

All these methods require specialised equipment, which adds to the cost of the deflection, and each has specific application. Mostly, they require multiple rod tripping, increasing time and cost. A single deflection by any of these would cost in the order of $A2000 to $A5000 (1991). One of the principal reasons why such methods might be preferred, is when the hole size or curvature prohibits 'natural wedging' (see Section 8.4). For example, the original hole above the required deflection point is already minimum size, and/or is so curved that devices such as chrome barrels will not pass down the hole.

Should an artificial wedge be used, then it should be done with careful planning and implementation; not just because of the cost but as it will affect the hole irrevocably.

The hardness of the wall rocks at the wedge point must be carefully considered. Very hard rocks, such as glassy rhyolite, may be harder than the wedge, and so it is the wedge which may be drilled out and not the adjacent rock. Very soft and talcy rocks may have the wedge pushed into the wall and the hole goes straight ahead.

Although the rod string will normally take 5° per 30 m quite well, it will not take it kindly if this is derived from repeatedly wedging. This is because it will not be a smooth curve, but a series of angular deflections which will induce rod fatigue as drilling continues. This can be alleviated to some extent by reaming out the sharp bend and so smoothing the curve.

8.1.1 *Hall-Rowe wedge (whipstock)*

This is a solid steel bar (see Figure 47), the same diameter as the rods, one side of which is machine tapered with the same radius of curvature as the hole diameter. It will have a chisel like blade at the base (thick end). The tapered end will have holes to allow a riveted connection (temporarily) to the rod string. The taper can be varied when ordering, but normally is 1.5° (e.g. 3.3 m *NQ* with 2.5 m tapered – active length).

A wooden block is inserted ahead of the wedge, into the hole, This should be softwood which will swell. It will be seated on the hole bottom or if required up-hole, will seat on a hole plug of sand, cement or mechanical device (e.g. Van Ruth Plug). When the wedge is then lowered on the rods, the chisel blade enters the wooden block wedging it against the hole walls, and thus hold the wedge so that it cannot rotate. As the wedge is pushed down by the rods, the rivets shear off, leaving the wedge securely in position and free of the rod string, if all goes well. If there is any doubt about the security of the wedge, it can be cemented, as a loose wedge can be a disaster later in the progress of the hole. Cementing of wedges is not always successful, however, as the cement finds difficulty in completely encasing, due to the small clearances.

A bullnosed bit is then run with a short solid reaming bar (about 40 cm) and drilled past the wedge 50 cm, then retract rods and fit coring bit and proceed. Although not recommended in medium to soft rocks, an old core barrel with an external stepfaced or worn impregnated bit will do the job, and so allow no hiatus in the core. Once the wedge is in the hole it will be necessary from then on to be slow when lowering rods past the wedge, as the bit can catch on the upper thin end: disaster if this gets bent over. When lowering rods it is best to use a bullnosed attachment to the innertube or a disposable carbon plug held by the corelifter, which protrudes from the bit. This is first to contact the wedge, well below the top edge, and then as it slides down it lifts the bit clear of the top of the wedge.

These wedges are non-retrievable, and all subsequent trips in and out of the hole involve running on the wedge, increasing the possibility of dislodgement. They are, however, relatively simple to use, and maintain hole diameter.

8.1.2 *Casing wedge*

This is basically similar to a Hall-Rowe, but as the wedge forms the last item on the casing string, plugs are not absolutely necessary, as the casing may be clamped at the collar to prevent dropping and rotation. These are usually 4.5 m long (3 m active length) so one gets 1.5° deflection for *NQ*. Whilst these wedges can be purchased, they are also easily made in the field from a length of casing, by cutting a tapered piece out, turning the off-cut over, and rewelding (Figure 42).

The procedure for use is similar to the Hall-Rowe except that the wedge remains attached to the casing string. The big advantage is that they are retrievable and often reusable, cheap, and quick to install. The main disadvantage is that the hole size is automatically reduced. Also, if one wishes to place thin PVC casing for geophysics for example, the steel casing would all have to remain in the hole. This because the PVC would be through the 'hole' in the casing wall.

Although like the Hall-Rowe, it must be tripped past subsequently, it is not nearly so prone to dislodgement or other problems.

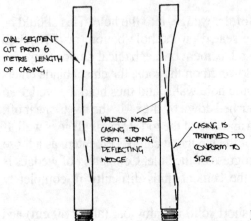

OVAL SEGMENT CUT FROM 6 METRE LENGTH OF CASING

WELDED INSIDE CASING TO FORM SLOPING DEFLECTING WEDGE

CASING IS TRIMMED TO CONFORM TO SIZE.

Figure 42. Casing wedge.

8.1.3 *Clappison wedge*

These are short length, two part reusable whipstock wedges which are designed to sit on the bottom of the hole. Once the new hole has been started, this time with an undersized bullnosed bit, the wedge is removed, and then it is reamed to full size. So, although some local reaming is necessary, they do enable the hole diameter to be maintained, but there is an interval of 1 to 3 m where no core is obtained. Being reusable they are ideal for keeping a hole on track, but are not good for starting a daughter hole from an old hole, although it is possible to do this if a plug is cemented in the hole for the wedge to sit on.

The following description of the device (see Figure 43), and how it works, is for the deflection of a *B* size hole. There are, of course, larger sizes available.

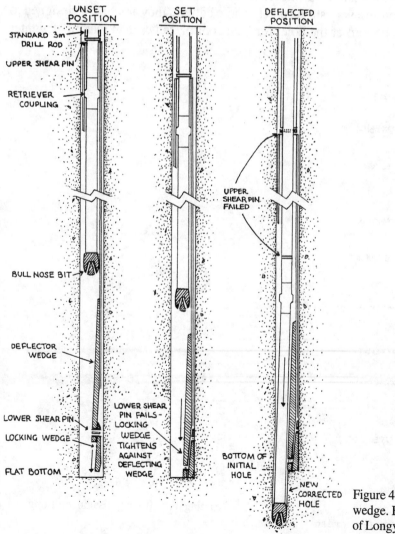

Figure 43. Clappison wedge. By permission of Longyear Australia.

The active length of the wedge is permanently secured to the bottom of a section of *BQ* sized rod. Attached by a shear pin is a small locking wedge, which, on contact with the bottom, shears the pin and allows the main wedge to slide down and tighten against the hole walls. The piece of adapted *BQ* rod is attached to an *AQ* reaming bar assembly by a stronger shear pin, which will shear after the wedge is seated and allow the *AX* bullnosed bit and reamer bar to commence rotation and advance to start a new hole off the wedge. After approximately 1.5 m of new pilot hole have been drilled, the rods are pulled, and firstly, the retriever coupling connects with the upper plugged end of the *BQ* sized wedge casing, which in turn allows the main wedge to pick up the locking wedge, and so all steel is retrieved from the hole. Now a reaming bar assembly (Figure 44), made up of a number of short pieces of bar with reamers and bits between to get maximum flexibility, is lowered, and the pilot hole reamed to full hole size.

As these require a very strong and flat hole bottom, they are mainly designed to be used at the bottom of the hole, for hole control, rather than making daughter

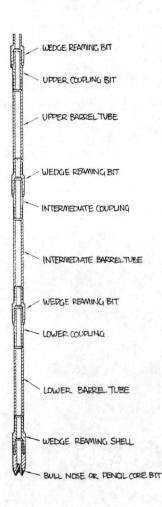

WEDGE REAMING BIT

UPPER COUPLING BIT

UPPER BARREL TUBE

WEDGE REAMING BIT

INTERMEDIATE COUPLING

INTERMEDIATE BARREL TUBE

WEDGE REAMING BIT

LOWER COUPLING

LOWER BARREL TUBE

WEDGE REAMING SHELL

BULL NOSE OR PENCIL CORE BIT

Figure 44. Reaming bar. By permission of Longyear Australia.

holes. These are not used very often then, as holes should be controlled with the rod string tools rather than wedges. An advantage they have is that the orientation, setting, and pilot hole drilling can all be done in one trip. Active length of *BQ* wedge is 71 cm, with base of 1.5 cm so we can get 1.2° (actually from 1° to 2° in practice). A similar device popular in Europe is Encore. In Canada the Thompson Wedge is a very similar device, which, by way of counter weights which tend to seek the low side of the hole, will automatically be orientated to deflect upward (the normal situation). These are inappropriate for holes steeper than 70°.

8.1.4 *Downhole motors*

These may be positive displacement motors (pdm) or turbines. The former are driven by fluid pressure and the latter by flow rate. Turbine motors cannot be made small enough for diamond drilling and the like.

Dyna-Drill, and Navi-Drill, the most common devices, use positive displacement motors to drill a range of hole sizes from 47.6 mm for diamond drilling to 660 mm for rotary drilling. They consist of an elastomer stator and steel helical rotor. Flushing water or mud passing through, turns the rotor. The drive shaft is connected to the rotor by a universal joint, and directional control is attained by either:
   – Orientated offset to the drive shaft at the universal joint housing or,
   – By a spring loaded deflection shoe at the lower bearing housing.
They are usually used without a core barrel, drilling with a barrel being continued after deflection is completed and pilot hole reamed. This means that each time a deflection is made there are several metres of hole not cored. The rear end allows downhole survey equipment to lock on, and so the device can be orientated. An electronic survey combination such as Well Nav, may be used to allow continuous survey with surface digital readout, as the deflection proceeds, (rods and barrel don't rotate). The point being surveyed is, however, always about 4 m or more behind the bit.

8.1.5 *Devibor*

In 1987 this was still under development by Craelius and was reported as follows by Suttill (1987).

'The devibor (deviation core barrel) consists of an outer tube with three eccentric bearings in which an inner tube rotates. The diamond core bit is attached to a spindle at the bottom of the outer tube. The spindle is driven by the inner tube. The rotation axis of the inner tube is set at a small angle against the outer tube to produce the deviation. The lower part of the inner tube also acts as core barrel for the 19.3 mm core.

Via left hand threaded casing, the outer tube is connected to a fixed chuck attached to the same cradle as the rotation unit on the drill. The inner tube is connected to the normal rotating chuck via drill rods. Both tubes are fed together, the outer tube being locked in position while the deflection takes place. During deflection drilling, the inner string is pulled out every 3 m to empty the core barrel. Once the calculated deflection has been achieved, the deviation string is removed and drilling continues with standard string.

The rate of deviation, which is always the resultant of forced deflection and natural deviation, can vary from 0.15°-1.0°/metre. The plane of deviation is changed by turning the outer string. An orientation instrument, in which the desired orientation has been preset according to the computer calculations, is lowered or pumped into position behind the upper bearing of the inner tube. The outer string is then turned until the instrument gives a signal indicating that the actual and desired orientations coincide'.

8.1.6 *Vic drill head*

The Norwegian company Devico A/S markets a very similar deflecting core barrel to the Devibor described above.

8.1.7 *Directional drillrod (ZBE)*

This device was designed to drill vertical holes only. The following description is again taken from Suttill (1987).

'The innovation of the *ZBE* is that it measures the smallest deviations continuously during drilling and counteracts at once.

Developed by Bergbau Forschung GmbH and Schwing Hydraulik GmbH, the present design has a diameter of 216 mm. The 1.95 m directional drill rod is installed behind the 216 mm drilling tool and can be used with standard pipes. It consists of an inner shaft and an outer non-rotating tube with four movable steering skids. Small chambers behind the steering skids house the inclinometers, steering electronics, magnetic valves and steering cylinders. Deviations from the vertical of as little as 0.06° are immediately corrected by hydraulic pressure on the appropriate steering skids.

The hydraulic and electrical energy for the steering components and measuring instruments are produced by the relative motion of the rotary shaft and the outer tube. Four hydraulic pumps provide 60 bar pressure, and a generator produces 45w at only 60 rpm. The entire device is flameproof.

Continuous monitoring of all functions is provided by an electro-hydraulic impulse transmitter built into the head of the directional drill rod. This passes timecoded pressure pulses for eight measurements via the flushing water column in the rod string to surface'.

In a test series of 17 holes from 72 to 258 m deep, in only one case did deviation exceed 0.3% of hole length. This device is undergoing further development for use in more vertical holes where it will be a very useful addition to the 'tool box'.

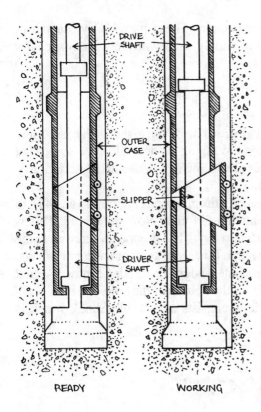

READY   WORKING   Figure 45. Continuous whipstock.

8.1.8 *Continuous whipstock*

This device, reported by He Yishang (1987) to have been used in China since 1981, is a device for deflecting the current hole at full diameter using a downhole motor drive. These will deflect 0.3° to 1.8° per m in *NQ* and *BQ* size.

The principle is illustrated in Figure 45. As thrust is applied, a slipper is pushed out against the wall to one side, which forces the bit to the other side of the hole. The slipper is fitted with rollers to allow it to ride down the hole in this position.

8.2 ORIENTATING WEDGES AND DOWNHOLE MOTORS

As the normal reason for using wedges and downhole motors is to change the direction and/or declination of a drill path it is necessary to position the device to obtain the desired result. The most common devices in use are single-shot cameras or acid tubes, (both described in Chapter 12), which are attached in such a way to the wedge that a reference direction can be obtained. Acid tubes are only used where the declination is between 25° and 80°, because outside these limits they are too inaccurate.

As commented upon earlier there are devices with continuous readout of digital data at surface by way of a cable connection. These may be gyroscopic or magnetic survey tools and are more commonly used with the more expensive downhole motors.

8.2.1 *Mule shoe*

The most common device to relate the camera or acid tube to the wedge is a 'mule shoe' (Figures 46 and 47). The different wedging devices all have somewhat different applications, but as an example the following is a detailed description of orientating a simple Hall-Rowe wedge.

At the upper end of the wedge is attached a dropper ring (Figures 46 and 47), to the inside of which is welded an orientating pin just below the thread. This is approximately 6 mm in diameter and 15 mm long. The dropper is screwed to the end of the drill rods, and fitted to the wedge by way of temporary copper rivets. The pin should be opposite the flat face of the wedge. The wedge is now lowered down the hole to a point about 0.5 m above the bottom of the hole.

The other part (the orientating device), consists of a 'Mule Shoe', which is

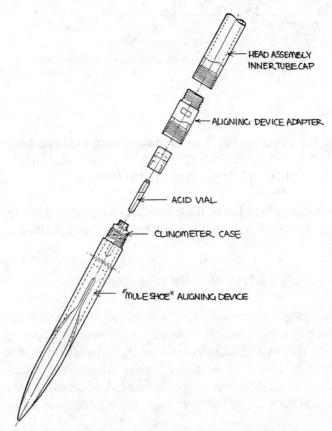

Figure 46. Mule shoe. By permission of Longyear Australia.

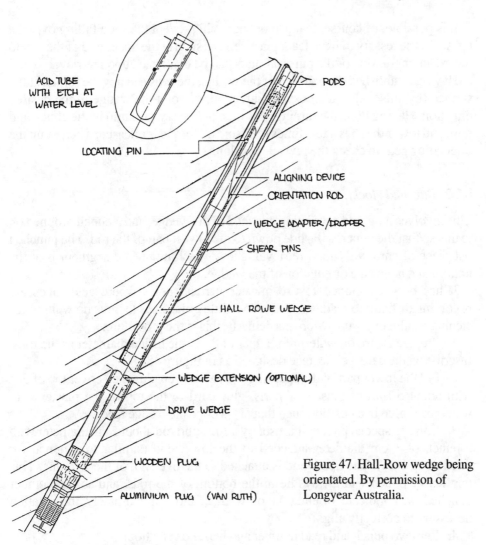

ACID TUBE
WITH ETCH AT
WATER LEVEL

RODS

LOCATING PIN

ALIGNING DEVICE

ORIENTATION ROD

WEDGE ADAPTER / DROPPER

SHEAR PINS

HALL ROWE WEDGE

WEDGE EXTENSION (OPTIONAL)

DRIVE WEDGE

WOODEN PLUG

ALUMINIUM PLUG (VAN RUTH)

Figure 47. Hall-Row wedge being
orientated. By permission of
Longyear Australia.

attached to the bottom of a normal acid tube (or single-shot camera case). The
shoe consists of a brass tube about 40 cm long with a lower end cut at about 30° to
the axis of the tube, and then, from the upper end of the obliquely cut piece, is a
slot approximately 10 mm wide and 20 cms long. When the shoe falls down the
hole it is guided by the oblique cut ends so that it can only proceed past the
orientating pin if the pin passes up the shoe slot. The angular relationship between
the wedge face and mule shoe is now fixed, and consequently, with the acid tube
or camera which is attached. Should a camera have to be used then the last three
rods (9 m) would have to be non-magnetic (austenitic stainless steel), and the
orientating tube and mule shoe must be brass with extension for 6 m in brass or
aluminium, and the orientating pin should also be brass.

It is possible, of course, that grit or such will prevent the shoe falling over the pin so it is necessary to have lead, plasticene or such at the upper end of the slot to record an impression of the pin so contact can be confirmed upon retrieval.

Having established the azimuth of the wedge, the rods can now be turned at the surface (by stilsons) in a clockwise direction, to point the wedge in the desired direction. Having done the turning, it is then necessary to clamp in the chuck and raise and lower the rods a few times to ensure all torque is released. Then rerun the orientating gear to check the revised position.

8.2.2 *Punched clockface*

This involves a figured card being attached to the wedge, and a punch dropped on to this so that the punch will slide down to the lowest part of the rod. The punched out piece of card will, upon retrieval, show which numbered segment is at the bottom and hence the orientation of the wedge.

It has been developed to use in near vertical holes where acid tubes are becoming difficult to read, and it is also a little quicker, as there is no waiting for etching or film exposure. More particularly, this is the procedure:

1. Prepare a card on waterproof paper of the same inside diameter as the rods. Inscribe on the card a clock face designed as in Figure 48.

2. Fill the lower part of dropper ring with a wooden plug then cover top of this with about 5 mm of plaster of paris, glue card to the top of the plaster with waterproof glue in a position such that '1' is opposite wedge face.

3. Lower a special punch. It is usually a 1 m solid rod about 2 cm diameter with a punch of 1 cm diameter attached to the lower end (similar to a punch for orientating core). The upper end is attached to the overshot of the wireline. This thin heavy punch will always lie in the bottom of the rods, and so the number punched will indicate orientation of wedge, and approximately the rotation necessary to correctly align.

4. Retrieve punch and read number and hence orientation.

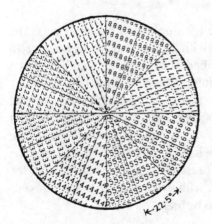

Figure 48. Clockface card.

8.3 PLACEMENT OF WEDGES

8.3.1 *Hall-Rowe wedge*

1. Prepare hole for the wedge – if wedge is to sit on bottom make sure it is clean, or at least free of fine cuttings.

If the wedge is to start elsewhere, then the hole must be blocked off at this point by either filling hole with sand; inserting a Van Ruth plug (mechanically expandable), or inserting a cement plug and cutting out to the correct depth.

2. Attach wooden plug to end of the rod string with a special adaptor and again lower rods into the hole and rest on top of Van Ruth plug, wait for wood to swell, then withdraw rods leaving wooden block.

3. Attach adaptor to the upper part of wedge with rivets. Normally these are copper, but if the hole is tight, or for some other reason one expects difficulty in getting down the hole with the wedge, steel rivets, or more rivets, can be used. This is not normally desirable, as the act of shearing off the adaptor may bend the top of the wedge into the hole and cause problems; e.g. top of wedge may then have to be cut away with bull-nosed bit to clear the way – this costs diamonds and can dislodge the wedge.

4. Lower wedge on drill rods to about 30 cm from bottom – must not touch bottom.

5. Using wireline, lower orienting device, and upon retrieval the present orientation of the wedge will be known.

6. Rotate rods by hand with stilsons, or such, to bring wedge to desired orientation. The rods should be rotated in the same direction as normal rotation to avoid unscrewing rods. Clamp chuck and raise and lower rods a few times to ensure the bottom of the string has rotated and is not held by any frictional torque.

7. Repeat (5) and (6) until satisfied with orientation.

8. Using hydraulics, force rods down firmly on the bottom which will firstly fix the wooden block and subsequently pin the wedge in position as the blade at the bottom splits the wood and wedges the block against the walls. Secondly, the downward force will shear the pins and release the wedge from the adaptor.

If there are any doubts as to the wedge being firmly placed one can cement back over the wedge before commencing drilling past. It is however difficult to get the cement to completely surround the wedge and hence hold it securely every time.

The wedge should not be placed in very broken ground such as a fault zone as the wooden block may not get a good grip. Also, very hard ground should be avoided as the rock is nearly as hard as steel and so the maximum deflection may not be obtained.

9. Withdraw rod string, and, using external step faced bit (5 or 7 steps) or bullnosed, drill ahead slowly.

8.3.2 *Casing wedge*

The procedure is the same as for a Hall-Rowe, except that here the dropper with orientating pin is part of the casing string and so remains in the hole, with the pin being drilled out as wedging commences.

8.3.3 *Clappison wedge*

1. Run flat faced noncore bit to clean hole and make a flat face.
2. Attach orientating rod with pin to normal rod string, followed by the complete wedge assembly.
3. Lower, orientate and set as for Hall-Rowe (but two set of rivets to shear).
4. Drill and back ream 15 cm. to taper the top of the wedge, and drill ahead 1.5 m, then back ream till the bullnose clears the top of the deflecting plate.
5. Remove wedge from hole (must be all parts).
6. Assemble and insert reaming bar and open the hole to full size, then remove.
7. Run normal rod string, preferably with a short barrel for a while, and carry on normal drilling.

8.4 NATURAL WEDGING

Natural wedging means making a daughter hole without using any of the specifically designed mechanical devices, other than normal rod string components. These methods make use of internally stepped or flat faced bits, larger diameter rod string, stiff barrels, reamer positioning, and slow penetration rates, to create a path on which the rock strength anisotrophy does not have a significant effect.

The most usual way these are used is in underlie wedging. The term 'underlie' is used here to describe those holes which are wedged in a direction opposing the natural deviation of the hole. The most common situations are drilling deeper from an existing hole to obtain a deeper intersection of the same mineralization as intersected in the existing hole, or endeavouring to maintain a straighter hole path than the hole will take without this help.

8.4.1 *Hole size continuation*

Commonly where drilling is being done in an anisotropic rock situation, and the hole has been reduced in diameter, e.g. *NQ* to *BQ*, then an increased curvature will result. The ideal place in such a hole for an underlie wedge, is at this diameter change. The *B* rod string should be removed and the *N* rod string re-equipped for coring; then drill ahead slowly with flat faced or internally stepped bit. Other features such as chrome barrels may be added. The new hole will diverge from the original, the new one being straighter. Obviously the cost of such a wedging operation is minimal, say $A200 (1990).

8.4.2 *Hole 'Lipping'*

This technique is suitable where it is required to deflect a hole in which there has been no size reduction, and nor is it desirable or possible to do so.

The objective here is to move the cutting head up and down over a constant length of hole for some time so that the under side (in normal situations) of the hole will be cut and eventually a 'lip' or ledge will appear where the cutting head is bottoming. Once a ledge of about 1 cm has been created one can drill ahead – slowly for a metre or so, then resume normal pressures, etc.

Further details on the execution of a 'lip' are as follow. Fit the string with an internal step bit and a chrome barrel if available. Select the precise place (± 0.1 m) where the branch hole is required which should be influenced by an increased rate of curvature in the original hole and softest rock. Commence drilling with slow feed rate (2 m/h maximum), at a point about 3 quill lengths (or 2 m for top drive rig), above the required take off depth. Drill forward the 3 quill lengths and repeat. Then withdraw bit 4 quill lengths (must be overlap), rechuck, and drill forward to the same spot, and repeat. Then withdraw bit once more for 5 lengths and repeat whole procedure. A ledge or lip should now have formed (Figure 49), and the hole can be continued, slowly at first, as should the ledge break away, then a branch hole will not be formed.

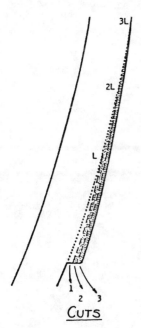

Figure 49. Hole lipping.

In soft rock once the ledge is produced, even rod weight could break it off, so one should not then test for the presence of a ledge – if it is going to work, one has to assume that it has formed. If there is any doubt, it is better to spend more time making certain of a ledge. The feed rate required to create a ledge will vary with rock hardness. Very hard rocks may only need 1 m/h or less.

CHAPTER 9

Selection of drilling method for exploration

Essentially the drill is a sampling device, and we will keep this in mind as we consider what type of drilling rig is required for various objectives.

Firstly we need to define our objectives and the principal controlling factors, so here is a checklist of items, many of which are discussed in detail elsewhere in this book.

Define target
- Geographical location – nearest town, etc.;
- Precise position – northing, easting, *RL*;
- Material – sulphide, alluvials, saprolite, coal, etc.;
- Attitude – what angle of attack necessary;
- Precision – how close to target must be achieved.

Define host rock
- Lithology – as reflects drillability;
- Structure – attitude of anisotropic features;
- Weathering – drillability and wall stability.

Define water presence
- Water table – drilling dry or wet; how far;
- Water flows – how much air required, etc.;
- Water supply – for drilling and camping.

Define sample required
- Quantity – metallurgy or assay or geochemistry;
- Accuracy – ore reserves or geochemistry;
- Disturbance – core or chips.

Define access
- Prospect – access for stores, etc.;
- Site – space for rig manoevering.

Define politics
  – Greenies;
  – Land occupier;
  – J.V. partners;
  – Management.

If we consider all these we will be able to make an informed decision as to the most suitable method. It will also enable the calling of tenders in a logical manner. We will now consider some hypothetical examples.

Example 1

Target. Near Charters Towers, at Youbeaut, 1005E 507N RL702 (surface is RL1000), massive sulphide, dip 80°N, strike 270°, precision required is ± 10 m.

Hostrock. Hanging wall is sericitic altered tuff, footwall is hard glassy rhyolite. Foliation present in hangingwall parallels ore zone. Weathered to 10 m, but will core and hold up well.

Water. Watertable at 20 m, major water flows unknown. Drill water available 500 m away in dam.

Sample. Assay for possible ore reserves only.

Access. All weather road to prospect, dozed near level tracks to cleared, level site large enough for truck.

Politics. No local 'green' action, grazier reasonable but no camping near dam. J.V. partners do not like precollared holes.

Now let us go through the selection process.

   Firstly we will need a machine to drill a declined hole of the order of 350 m at least. Because the hole is declined, a cable drill is definitely not suitable. With a hole depth of 350 m, an auger or uphole hammer drill are out of the question.

   As a precision of ± 10 m would be very difficult at this depth, with the rotary and downhole hammer drills, we are left with a diamond drill, which would easily achieve this. We now make a final check of other features, and make sure there are no important negatives, apart from cost, which must be a final consideration, if more than one method could be used.

   We could therefore plan on multipurpose drill to drill downhole hammer precollar to 100 m at least, (depends how it stays on target – this may satisfy J.V. partner who would rather have a cheap intersection than an accurate one). Then complete with preferably *NQ* core, but with the possibility of *BQ* if control becomes difficult or if a smaller sample is adequate and cost saving required. A multipurpose rig would best do the job.

Example 2

Target. Near Newcastle, at Hunter Downs 103E 8776N RL023 (surface is RL151), coal, dip 5°, strike 290°, precision required is ± 0.1 m in RL and ± 5 m in the horizontal plane.

Hostrock. Soft sandstone and shales, bedding dip 5°. Weathered to 5 m, but will hold up well.

Water. Watertable at 10 m, moderate water flows will be encountered from 60 to 80 m from surface.

Sample. Totally undisturbed, assay for possible ore reserves.

Access. All weather road to prospect and dozed tracks to cleared level sites large enough for a truck.

Politics. Possible 'green' action, farmer has been compensated, management wants cheapest alternative.

Now the selection process.

This time we will need a vertical hole drilled to about 150 m. This is too deep for an auger, and because we must have an undisturbed sample, this must be core, and preferably from a triple tube, due to the very variable strength of the coal and hostrocks. We certainly don't need to core until about 100 m, and maybe further. This part of the hole could be done rotary with air, and finally water or mud, or it could be a downhole hammer. An uphole hammer could not be used because of the water.

A multipurpose rig would certainly be used, because with the relatively short hole it would be costly to change rigs, and the desired precision should be achieved with a conventional truck mounted rig.

Example 3

Target. Near Ravenshoe, at Cow Bell Creek, along creek bank, alluvial, precision required ± 0.1 m vertically and ± 1.0 m horizontally. Depth 10 m.

Hostrock. Gravel and sand to 150 mm diameter with clay bands.

Water. Watertable at 1 m and weak to moderate flows may be expected at any depth.

Sample. May be disturbed, but must have good prevention of contamination from walls, and must have good depth control – for ore reserves – reconnaissance completed.

Access. All weather road to prospect and dozed tracks to cleared sites, some of which are too small for a truck.

Politics. Possible 'green' action, State forest, management wants the most reliable sampling available.

Now the selection process.

Vertical holes will be required, so all techniques qualify. The poorly consolidated nature of the gravels mean that both mechanical hammering devices are not very suitable. Ordinary rotary drilling will also suffer from wall contamination. Reverse circulation rotary drilling would solve this, but the size of the boulders to be either broken or flushed, makes this also not the best. Most screw augers available will not handle this size boulder or, if they have a large enough diameter, they will not handle the water flows, with subsequent contamination.

We are left therefore, with either cable percussion or bucket auger, both of which would be satisfactory if casing follows the bit. However, as the management wants the best, then at probably twice the drilling cost and up to 50 times the sample preparation cost (because of the much larger sample volume, with say a 900 mm hole compared to a 200 mm hole), the bucket will give a 'Rolls Royce' result. Sampling (usually pilot plant) will have to be done elsewhere to satisfy environmental aspects, so trucks will have to cart the cuttings and return fill for the holes.

Example 4

Target. In Waihi, New Zealand, urban area. Sub-vertical gold bearing quartz veins. Precision required is $\pm$ 5 m horizontally and $\pm$ 2 m vertically. Depth 120-300 m.

Hostrock. Andesitic volcanics overlain by 50-100 m of soft fractured and oxidised post mineral ignimbrite.

Water. Watertable at 10 m – no drilling problem. Drilling water obtained from town supply 300 m away.

Sample. Large core for assay for possible ore reserve.

Access. All weather road – site already cleared.

Politics. Must keep noise to minimum. No drilling 10 pm to 7 am. Rig to be as small and unobtrusive as possible.

Now the selection process.

We will require declined holes to 400 m maximum to complete at *HQ* or larger. The accuracy required confirms diamond coring as best type of drill. As no core will be required through the ignimbrite, the early part of each hole does not require core sampling. Because of the poor ground conditions there would only be a choice of air circulation with rotary or hammer which would be noisy and would involve a very visible machine. Using a conventional diamond drill and precollaring *HW* with a rotary bit fitted with casing advancer gives a (relatively) quiet, low cost and low visibility operation.

CHAPTER 10

Preparation for drilling programme

10.1 PRELIMINARY PLANNING

We are now at the stage where we have conceived drill targets, and have considered the best method to achieve the objectives. We now need to record the intention so that:
 – You can review the programme as it is fleshed out;
 – Your supervisor can review the programme;
 – Potential tenderers will have an adequate database.
The following schedule is suggested:
  1. Obtain or purchase suitable plans of the area. Show drill sites in relation to: Access; Water supplies (drilling & potable); Stores supply (fuels, food, etc.); Regional geology.
  2. Show on both plan and section at a large scale, the approximate hole collars and targets, and preferred drilling sequence. It is usually best to use other names than the expected final hole names as often the holes are actually drilled out of sequence, or drilled in different locations, which may mean a change of name. For example if holes are named in order of being drilled, and have numerical names such as TH45, TH46 etc., then proposed holes could be called A, B, etc.
  3. On detailed geological plan and sections, plot the proposed drill hole paths, and show hole sizes desired. (How to predict a hole path is discussed in detail elsewhere.) Geological data inferred, should be shown clearly from that which is factual.
  4. Write a brief report stating the proposed programme, the type and size of rig to be used, the reasons for it, the expected results, and the estimated costs (contractual, supervision, assaying, administrative etc.).
  5. Submit all this data for approval, or if at a senior level, at least hand it to someone to check. It only takes an hour or so to check a $25 000 hole.

10.2 CALLING DRILLING TENDERS

Prepare a brief statement of requirements and conditions, which, together with the access map referred to under preliminary planning, can be sent to those contractors whom you would prefer. There is no point in asking every contractor to quote, as there may be some with special experience or a better suited rig or they have been working on the same project before. The statement should contain the following information:

1. Distance to nearest major highway and railway and town from which fuel and stores will probably be available.

2. Distance and road conditions, and perhaps elevation difference from sites to drilling and potable water supplies. The water source should be stated e.g. from a shaft (metres to water), from a bore (capacity and depth to water), from a creek, etc. Make sure permissions to use water are available. The quality of water should be stated.

3. Number of holes and total metreage, in two parts, if possible: 1) Guaranteed; 2) Expected.

4. Hole depths: 1) Minimum; 2) Expected; 3) Maximum.

5. Hole declinations.

6. Size of core required, particularly at target.

7. Depth of overburden and oxidation – are precollars required.

8. Terrain of drill sites and whether preparation will be contractor's responsibility; how access will be made, and if by helicopter, weight of components allowable.

9. Available accommodation.

10. Required starting date and progress rate.

11. Drilling technique to be used, e.g. *RC*, diamond core, etc. or combinations thereof.

12. What manpower is preferred, i.e. do you want 7 day/24 hour activity or 5 day/8 hour activity, or something in between such as 7 day/20 hour or 7 day/10 hour. I have a personal preference for activity to be as continuous as possible, which reduces supervision time, and therefore costs.

13. Any other pertinent fact which will affect the drilling, e.g. rock types, prior drilling history, wall rock stability, water loss, etc.

14. Should the holes to be drilled require any type of hole control, then it is best that the contractor have a 'toolbox' of critical gear available on site or no more than 4 hours away. The following may be considered essential for high accuracy deep drilling: 1) Internal step surface set bit; 2) External step surface set bit; 3) Chrome barrel; 4) Casing wedge; 5) Hall-Rowe or Clappison wedge.

15. Who is responsible for delays due to weather and to what extent. i.e. what does it cost whom.

16. Who supplies consumables such as PVC collars, fluids, etc.

17. Type and accuracy of any downhole survey instrument required.

10.3 SELECTION OF CONTRACTOR

Tenders, when received, are often in very different formats, and so one has to apply each set of costs to a simulated programme to enable realistic comparisons to be made. In costing such simulations, one must remember to consider the peripheral costs such as geological supervision (see preferred manpower in previous section), delays, material used or lost down hole such as casing, the quantity and type of additives used (some drillers unnecessarily 'Rolls Royce' this, as it is direct cost to client).

The contractor selected should have had experience in this particular type of drilling, and preferably have experience in the area. For example, one would not normally select a contractor usually drilling coal at Newcastle to drill iron ore in West Australia. Also, like Napoleon experienced, the lines of communication would become stretched.

Be careful of exceptionally low prices. This could mean that the equipment is inferior, the contractor is in financial difficulty, or maybe it is just an error. Maybe on the other hand it is as one contractor confided in me, 'first get the contract and then screw the principal'. Whatever the case, on the basis that happy contractors and drillers perform best, it is not in your interest to have a contractor not earning reasonable money.

Beware of the one-rig contractor if time is important, as any man or equipment failure will not be so quickly rectified as by a larger organisation. There are, however, many very efficient owner operated rigs who you will get to know by reputation, and there are many jobs for which they are ideal, particularly those with large distances between drill sites and difficult accommodation and access. The deeper the holes and the more the total metreage involved, the fewer contractors will be considered; but then, because of the higher total cost involved, these must be given even more thought.

There are people who, before calling tenders, have already decided who will get the contract. This may be for good reasons such as a contractor has done all work on site to date, and you and/or your staff have had good relations.

It is not really ethical to have other parties put their time and effort into the tendering process if they have no chance of success.

It is also not ethical to hold 'Dutch auctions', where quotations are received, and then informing some of the tendering parties what the others have quoted to enable them to requote at equal or lower prices. This is effectively bidding the price down in a covert manner.

Most important, of course, are the people who will man the rig and their experience. This may be more important than who owns the rig. Whilst few contracting organisations will allow a client to select which of their drillers will appear on the job, I believe it is reasonable for a prospective client to ask which staff will be used, should the tender be accepted.

So on the basis of price, reliability, experience, available machinery, quality of drillers, a contractor is selected and contracts are exchanged. These contracts must provide legal answers to as many eventualities as possible. Copies of the contract must be available on site, and not just at head office or arrive on site as the drilling rig departs. Appendix 2 contains a sample of a brief drilling contract agreement, which could provide a framework to make up one suitable for most situations. Appendix 3 contains a suggested appendix to any drilling contract in New Zealand which will ensure that the rather strict environmental legislation is complied with. Modification may make much of this applicable to situations elsewhere in the world as well.

10.4 PREDRILLING PREPARATIONS

There are a number of things to be done by the site geologist before the rig arrives. In many contracts, some of these items will be the contractor's responsibility, but, in order that nothing will be overlooked, let it be assumed that the geologist has the responsibility, and whilst the following items refer specifically to diamond drilling, most are applicable to other methods.

1. Arrange purchase and delivery of core trays. Allow 30% excess in all sizes.
2. Arrange to have, or to be able to readily obtain, a core splitting device; preferably a diamond saw.
3. Arrange core depth marking blocks, sample bags, logging stationery and chemicals.
4. Inform land owners, lessors or occupiers.
5. Arrange a site for core logging and storage, so that animals and possibly people cannot disturb.
6. Arrange access to water and transport of same (truck or pump and pipe).
7. Arrange with your supervisor sampling techniques, sampling intervals, assay specifications and preferred laboratory.
8. Arrange camping or other accommodation.
9. Arrange personnel for logging and core cutting, etc.
10. Prepare drill sites and access. Locate or construct an unloading ramp or some alternative unloading machinery, if required.
11. Arrange downhole survey equipment and associated spares, such as batteries, developer, acid, etc. if drillers are not to supply these.
12. Arrange sample preparation equipment, if required, such as splitter, bags, etc.
13. Arrange photographic gear to photograph core, if required.
14. Arrange down hole logging, if required.
15. Arrange permanent core storage.
16. Inform any other 'in house' department that could be interested.
17. Inform the Mines Department of those details required by law.

18. Mark out drill sites (or at least the first).

19. Should orientated core be required, then ensure that the contractor will have the necessary equipment to get this, and that the necessary measuring devices are available to assist logging.

Some of these items require further elaboration.

10.5 SITE PREPARATION AND MARKING

No matter which rig is used, except perhaps a track mounted percussion rig, a reasonably level site is required. Ideally a 15 m square area is adequate. Keeping in mind a responsibility to eventually restore the area to its former natural glory, all grass and small vegetation will be cleared (especially in dry areas where there may be a fire hazard), and trim larger trees, branches and shrubs which interfere with the practical operation of the rig. Special rules may apply in scenic areas, Environmental Parks, World Heritage Areas, and places like New Zealand, where much more restrictive limits are in force. Powerlines should be identified.

A sludge pit or holding tank should be dug or provided for diamond drill holes. This should be positioned downslope from the collar. The edge of this pit should be about 6 m from the collar, and it should have a capacity of at least 500 l, but preferably 5000 l (rule of thumb ideal size is 5 times the volume of the hole).

If the rig is not truck mounted, and is on light skids or such, as is very common in a helicopter supported operation, it will require holding down against the drilling thrust, and, if the mast is not an integral part of the rig, then against the weight of rods to be pulled. There are several common ways of doing this.

Bed logs
Two logs of the order of 2 m long by 300 mm thick are laid in 1 m deep trenches directly under where the main frame of the rig will be. Wire ropes are then fastened around each end of each log and the trenches backfilled leaving the four wire 'tails' reaching out of the ground with sufficient length to be shackled to the rig frame and pulled down with turnbuckles.

Anchor bolts
Drill four holes with a jackhammer or such, then cement in 1 m long eye bolts to which the rig frame can be attached as above.

Anchor rod
Cut some crude teeth on the end of an old drill rod and drill this into the ground without fluids, and it will eventually 'burn in', which is to say frictional heat will allow the end to grossly distort and become immobile. One such anchor may be enough if it were placed approximately under the main winch and drill head.

The marking out may seem quite simple but it is rarely done with the care it deserves.

Firstly check compass with grid direction and note any correction factor. If grid exists, measure from nearest grid pegs, and then check position with bearing and distance to another peg. Then check bearing and distance relative to target's surface expression e.g. gossan, geochemical or geophysical anomaly, etc. Finally get another person to check at least approximately. All this may seem a lot of checking but with the cost of an average diamond drill hole at about $25 000, an hour or two checking is warranted. I have seen holes drilled 100 m from intended collar, and 180° from intended azimuth because of bad marking out.

So now that the correct collar position has been established, this should be marked with a short peg about 150 mm high so that a rig may later drive over without disturbing it. Somewhat longer (up to 1 m) pegs are then driven along the intended azimuth of the hole and situated about 25 m in front and behind. Similar sized pegs should be driven to the left and right about 10 m. These sighter and side pegs will therefore allow the collar position to be recovered should the collar peg be disturbed.

The collar peg, or an auxiliary peg next to it, should clearly be marked with hole number (name), azimuth and declination. The sighter pegs should be so marked e.g. *FS & BS*. All site pegs should be painted or flagged very differently to existing grid pegs.

CHAPTER 11

Managing the drilling programme

11.1 SUPERVISION OF CONTRACT

The supervision of any contract requires a supervisor to act with great professionalism and, although representing the principal, he or she is expected to act with honesty and impartiality in respect of any dispute. The supervisor, as the principal's representative, must of course ensure that the contractor or his representatives, perform according to the contract. More importantly, the following particular duties must be attended to.

11.2 SUPERVISION OF COMMENCEMENT

1. Arrive at site at the time the rig is due.
2. Show the driller or foreman, where all sites are approximately, location of water supplies, camp area, etc, so that unloading may be done in the most suitable locations.
3. Provide driller with detailed instructions on the first hole. Whilst there will be further matters to be discussed with the driller under the banner of hole control the following are the more basic items to be communicated:
   – Exact drill collar location;
   – Core sizes required;
   – Depth to start coring;
   – Target depth;
   – Collar declination;
   – Proposed drill path with drilling techniques expected;
   – Expected rock types;
   – Expected sub-surface hazards (Figure 32);
   – Casing required to remain in hole;
   – What downhole survey procedures will be used;
   – Communication procedures;
   – Who has authority to be on site and their position;

– Who is land occupier and how he or she should be treated;
– Any special environmental factors;
– Where and how to dispose of rubbish and/or drill sludge, etc.

4. Line up the rig. This is a task that most geologists, and many drillers, don't know enough about. Under ideal circumstances, in which the rig is in its original, as manufactured condition, the drill head should rotate at right angles to the main frame. In such cases it would appear satisfactory to line up the frame or truck body with the required direction. This would be true if the rig was new and if the frame was horizontal. Yes, we have all seen the driller use that old 400 mm spirit level, and sit it on a bent, welded, dented, or covered in grease, part of the frame, and of course, this would be of doubtful accuracy. It is with this in mind, that I strongly recommend the following method.

a) Approximately position the rig and make it approximately level (by eye will do) and line up direction using frame or truck body.

b) Insert 6 m of rod in chuck and elevate mast to approximate declination (i.e. ± 2°).

c) Thread string line through the rods, and hang 'plumb-bobs' on each end to almost reach the ground. The direction between the two plumbed strings is the direction in which the machine will drill a hole.

d) Now there are two equally valid ways of making the precise alignment, either push, pull, or steer the frame to get the correct direction, or what may be much easier, particularly with jack-up bases, just tilt a little off level until the direction is correct.

e) Make final adjustment to declination. Now the rig frame may not itself be lined up, but the drilling direction will be.

Many truck mounted topdrive rigs do not allow space for this procedure, but on the other hand the drilling direction is more likely to be parallel with the frame. For holes requiring accurate placement, it is better to use skid mounted rigs or use a theodolite to align by calculating azimuth from positions at bottom and top of 6 m of rod in the mast ready to drill and proceed as in 'd' & 'e' above. Remember the more vertical holes require more care.

11.3 SUPERVISION OF DRILLING

The amount of on-site supervision will depend on the drilled metreage per day, number of shifts, quality of manpower and ease of hole control (does it follow planned path).

Maximise time at rig, with at least one visit per day. Preferably log core on site.

Communicate with drillers to be aware of any difficulties, and how they are being, or were overcome. Keep informed as to what gear is on the rod string, and what additives are in use, and why.

It is appropriate here to reflect on the cost of additives and their use.

Normally, an exploration company has a fixed budget, and an expectation of what percentage of that will be devoted to drilling. This means that the driller's maximum cash flow from any one client is fixed if they get all that client's work.

A normal exploration drilling contract is made up of two parts:

1. Metreage rate which has an inbuilt profit margin of the order of 20%;

2. Additives etc. at cost, plus a percentage to cover handling costs – usually 10%.

It is obvious then that if item 2 costs can be contained, there will be more funds available for item 1, and hence an increase in total profit derived from the client. The client will of course be happy with more metres for his money.

So let's look at a hypothetical example.

Budget $1 000 000 of which $500 000 is allowed for payments to contractors.

If drilling costs are $65/m and extras average 15% more, then total metreage cost is $74.75/m.

Contractor gets a profit of $86 957. Client gets 6689 m.

But if we reduce extras to 7.5%, the total metreage cost will be $69.87/m.

Contractor gets profit of $93,028. Client gets 7156 m.

This can only be done where there has been overuse of additives in the first place. Because additives are cost plus, there is usually not the level of supervision by the drilling supervisor that applies to other activities, and the geologist, in ignorance of the use or quantities, usually goes along with the driller, who often has a philosophy that if a little additive works well, then more will work better.

So if geologists can have a better appreciation of how additives work, then they can sensibly discuss with drillers the pros and cons of usage in any situation.

Let me quote two examples where the drilling has benefited from geologists knowing a bit about it.

Cyprus Gold New Zealand decided they should only use mud as a last resort instead of in every hole as was being done. Their contractors were supplying the additives. They now stock all their own additives and supply to the drillers only when they are convinced it is necessary. Now drilling is significantly cheaper and drillers get more metreage per shift, and the sort of cost arithmetic I was describing earlier is in place. The previous cost of consumables was 20% of total invoiced cost (meterage, field cost hourly rate, and consumables), it is now 3% on the same basis.

I recall that at Thalanga (Queensland, Australia), there was one occasion where I had to insist on mud conditioning of hole walls through sandy overburden prior to casing, to ensure casing removal upon hole completion. My local knowledge was based on watching previous drillers and how they solved their problems. The saving was hundreds of metres of recoverable casing.

In this way the site geologist acts as the local library, and can provide useful historical data, particularly if there has been a change of driller or contractor.

If the drillers are doing the downhole surveys, either do a check survey yourself now and again or be present when the drillers do them – it is easy to fake surveys, though I have never seen it happen yet.

Make sure that the core does not get greasy. There are three major sources:
1. Greasy hands;
2. Excess grease in barrel rear end;
3. Rod grease from previous hole where current rods were used as casing, or from the circulating water drawn in as the innertube is pulled.

The evidence for, and remedy for the first is obvious. The second will show as grease on the core at the start of each run and the remedy is again obvious. The third will show as patchy grease (rod grease) all through the run. The remedy is to pump some diesel through the rods, if allowed by environmental constraints.

Core with grease should be cleaned immediately:
– Light grease – wash with soluble oil;
– Moderate grease – use degreasing oil;
– Heavy grease – as a last resort, wash in diesel or petrol, but then immediately wash with detergent or washing powder to remove diesel or petrol, which if it soaks in, will be just as bad.

Greasy core, or even diesel washed core may well be useless for any metallurgical tests.

Encourage the driller's offsider to pack core in trays neatly, but not too tight. Impress on him why the core must be in the correct sequence and attitude. As well as marking the metreage on coreblocks, also mark metreage on the tray base so that if blocks are moved, they can be so replaced.

Occasionally, and especially on night shift, small pieces of core are dropped on the ground, unnoticed at the time. When they are later observed the offsiders usually either put it in the tray 'somewhere' or they are thrown away. There are two ways of checking the second 'crime', firstly by walking about the rig at about the distance of a casual throw or secondly check the sump at the end of the hole. The offsider should be encouraged to collect such scraps and place them in a scrap tin so that the geologist can have the option of replacing them in the tray.

Sometimes there is a core souveniring problem. This should be discouraged – best to give away a piece of less important core.

At the end of each hole, count rods in or out of hole, to verify depth.

Whilst the contractor is normally expected to provide clean core, when using mud etc. it is often best to clean it yourself. In the case of soft powdery material, such as when sooty chalcocite is present, for example, the chalcocite may only be held by the mudcake on the outside.

Should 'controlled drilling' (i.e. less bit pressure than for normal cutting) be necessary, and it should be a last resort (see hole control), then almost constant supervision will be necessary. Drillers will often do the controlled metre limit in part of the shift, and do no drilling for the remainder. Remember that it is bit pressure that is important, not metres per shift.

Table 13. Site hazards.

– Power cables, steam or high pressure water and air lines.
– Ground stability e.g. landslides.
– Poor lay-out, difficult escape route and site mobility.
– Poorly stacked rods, etc.
– Loose overhead objects in mast, etc.
– Rubbish about site.
– Muddy or slippery walkways, ladders, etc.
– Potential flood area.
– Vehicle access track in work area.
– Open sumps.
– Bush fires.
– Ventilation.

Ensure that drillers and yourself obey the Mines Regulations Act (or equivalent), particularly with regard to personnel safety, and that the site hazards in Table 13 are addressed.

Bring to the attention of the foreman, or driller, any core losses or mistakes in depth marking, as soon as possible. In conjunction with him, endeavour to form an opinion as to why core was lost or apparently lost and rectify as necessary the depth markers.

Ensure that the driller's logs are adequate and are approved or otherwise on a daily basis. Such items as metres drilled, at what size, what casing was used, what additives were used, and what time was used for different activities, are usually quite well documented. There are some items which are not usually so well documented, as these are not cost related, and so there is no incentive for a drilling contractor to encourage the recording of same. These are, however, best done by the driller as he is the one who is observing the total hole progress. The type of items I mean are complete details of depth, type, and size of wedges and plugs; detail estimates of water flows or losses at what depth; when core is lost – where did it occur, and why, bit changes and serial number.

11.4 SUPERVISION OF COMPLETION

Once the target zone has been intersected, make sure that an adequate section of footwall has been drilled. I suggest at least 20 m beyond significant mineralization is useful at the exploration phase. Many projects have holes which ended in 'ore', when the assays were available. In any case, footwall conditions will be useful data for mine planning engineers.

Make sure that sufficient downhole surveys have been done – this is the last chance.

Decide what casing will remain in the hole. If the hole is to be probed or extended in the future, the top outer casing must remain, although if the walls are in reasonable condition, it can be withdrawn to remove the shoe bit (if present) and replaced.

If steel casing cannot be pulled or jarred, it may have to be unscrewed or cut with casing cutters or blasted to part the string. The amount of time spent trying to retrieve casing, etc., should be about half the time, in dollar terms, as the value of the materials being recovered.

Insert slim PVC casing required for any later probe work, if required. This could be glued into 200 m lengths the day before to expedite insertion. This PVC must not be old stock with any significant ultraviolet exposure. Storage in outside areas will cause this to be brittle.

When all gear is out of the hole, it should be plugged or capped, and the collar cemented. A steel star picket should be placed nearby, and metal tagged with hole number, and perhaps azimuth, declination, depth, etc. Such a peg, apart from making the finding of the hole easier, will help prevent vehicles disturbing the collar.

Restore drill site area as environmental and legal considerations may require. Ensure all Mines Department regulations have been adhered to.

Sign all driller's logs and if these are not adequate to completely record the history of the holes then interrogate the driller before he leaves the job. A record must be kept of the current state of the holes, by which I mean cased to where, by what and what downhole obstacles there are.

The following is a list of non-contractural items which the site geologist must address at the completion, in more or less this order.

1. Complete geological logs.
2. Complete core photography.
3. Record all survey data.
4. Reduce all survey data.
5. Orientate core and log structure.
6. Do any geotechnical logging.
7. Do any whole core S.G. measurements.
8. Do any geophysical core logging.
9. Mark out intersections for sampling.
10. Split core for assay.
11. Store core.
12. Complete plans and sections.
13. Bring model up to date.
14. Write report.

11.5 HOLE CONTROL

By hole control we mean the application of what is known about hole deviation so that the hole will proceed along a planned course with minimal recourse to artificial wedges and the like. Primarily, this means the selection and or changing tools such as bits and barrels, to ensure that a hole follows a planned path with minimal interruption to metreage progress.

Once tools have been changed, you will rarely notice an immediate effect of great magnitude, and it may take 30 m or more to really notice the effects. It is therefore essential to carefully monitor the progress, keeping in mind the tools, and trying to anticipate requirements. Don't panic if small variations occur – it is like steering a vehicle, you can overdo the correcting. Also remember that every bend in the hole gives greater drag, requiring more power to rotate, and therefore slower drilling (& greater costs). In addition, the options on tools to be used tend to get less as more are used, as some are not compatible, e.g. cannot get a chrome barrel down a hole with a very sharp deviation.

Hole control should not be confused with 'controlled drilling' which is controlling the penetration rate or bit weight.

11.6 HOLE CONTROL TOOLS

11.6.1 *To reduce deviation rate*

1. Use oversize core barrels (commonly chrome plated to reduce wear and vibration). These reduce hole clearance, are much stiffer and hence have longer active length. A disadvantage is that in very soft ground (where they are most needed), the penetration rate is reduced, due to reduction in cutting sludge dispersal from the bit due to less wall clearance.

2. Flat faced or internally stepped bits in anisotropic rocks will reduce the 'wedging effect' of strata layers.

3. Larger diameter holes will have larger rods which are stiffer. In rotary and downhole hammer rigs a large diameter stabiliser bar (either a complete rod of increased diameter, or spacers added to increase the diameter in places, is inserted above the bit (Figure 50).

4. Use longer core barrels giving a longer active length.

5. Use backend reamer only, which reduces hole clearance at the barrel whilst still giving good clearance for the remainder of the rod string, and hence does not materially increase the torque, but increases the active length. Rotary and down-hole hammer rigs can use rod collars on the bottom few rods (Figure 50).

6. Rate of penetration can be reduced, or in other words, the bit pressure can be reduced. This is only applicable in moderately soft rocks, as a critical pressure must be applied to avoid diamond polish.

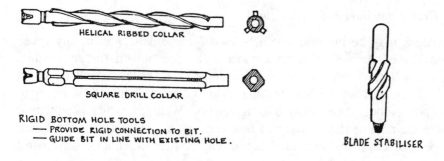

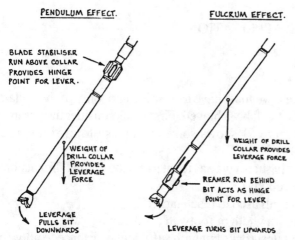

Figure 50. Stabilisers and rod collars. Reproduced from the *Australian drillers guide* by permission of the Australian Drilling Industry Training Committee.

11.6.2 *To increase deviation rate*

Increase hole clearance and decrease the active length by means such as new bits, external stepped bits, worn short barrel, oversized front reamer, no backend reamer and more bit pressure.

11.6.3 *To alter azimuth deviation*

It is difficult to alter this, independent of the declination, as generally the same principles apply. The only technique other than wedging, is to change the direction of rotation. This may work, although I have no experience of this, but it is said to make a normal declined hole drift to the left. There would be the need for a rare left hand threaded rod string and accessories, and there is also the problem of low revolutions available from reverse gear, though this would not affect rotation driven by hydraulics.

11.6.4 *Create daughter holes (wedging)*

Steel wedges may be inserted into a hole to force a deflection in any given direction. There are a number of devices available, some of which are retrievable. A daughter hole can be created in relatively soft rock by drilling ahead slowly at the desired point. Details of these techniques are given in Chapter 8. Wedging to deflect a hole is much easier than to straighten one. Wedging, which is only used to change the path of the operational hole (i.e. not to create a daughter hole), should be avoided, if possible, and should only be used when all other control parameters are not applicable.

11.7 HOLE CONTROL APPLICATION

11.7.1 *Hole planning*

Before a hole commences, we must have a clear idea not only of the collar, target and path, but also how that will be achieved. First let us consider the parameters of a well designed hole. Figure 51 shows some examples of how important good hole planning will be.

  – Intersects the target in such a way as to intersect the correct dimension (i.e. at right angles to a planar orebody).
  – Shortest possible length used to reach target.
  – Intersects the optimum width of footwall and hangingwall.
  – Drill path is in the softest rocks but with sufficient resistance to allow control measures. In effect the rocks are less costly to drill, and control measures have quicker and greater effect.
  – Avoids broken ground.
  – Provides good site access.
  – Provides good water access.
  – Avoids deliberate use of artificial wedging, and continuous wedging devices.
  – Azimuth is aligned slightly to the left of bedding and/or foliation azimuth (allows for rotational deviation).
  – Considers the effect of bedding and foliation, in particular upon declination.
  – At the end of the hole, the hole should be steep enough to avoid the use of roller or pump down overshot if possible (i.e. greater than 35°). At about 30°, due to friction, the overshot will not slide down with sufficient momentum to latch the innertube and so rollers can reduce friction. If fitted with special washers it can be pumped down. This is much slower than normal gravity fall.

Having considered all these things, we now can start with the specifics. The following are some of the more important things to be done:

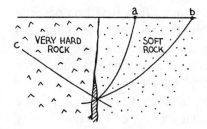

a. Shortest length, obtuse angle of intersection, in softest rock, NQ-BQ, allows easy wedge for deeper intersection, softer rock.

b. Longer length, obtuse angle of intersection, softer rock, NQ all the way.

c. Longest length to get the same angle, hard rock, BQ or NQ.

1. SECTION Vertical Orezone/ vertical foliation/same strike

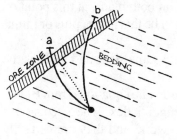

a. Allows for azimuth change due to bedding, therefore shortest length, and near right angle intersection.

b. Hole commences aimed directly at target with no consideration for deviation due bedding, longer and poor intersection angle.

2. PLAN Vertical Orezone/ vertical bedding/ different strike

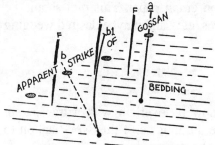

a. Targeted at gossan outcrop with path at right angles to bedding and so hole does not deviate.

b. Targeted with azimuth at right angles to line of apparent strike of individual gossan outcrops. Bedding deflects hole to path b1 and so misses target.

3. PLAN Vertical Orezone/ vertical beds/ transverse faults disguise direction of principal planar feature.

Figure 51. Examples of good and bad hole design.

a) Draw a working cross section (an approximate version may have been adequate for drillhole proposal and approval) along the planned hole azimuth, and through the target. Add the surface profile and probable geological features (a scale of 1:500 is usually most useful). Draw this on a sheet of gridded (1 mm) semi-transparent paper preferably.

b) One should now work from the target, back to the surface to arrive at a hole path. So, at the target, draw the intercept you would like to make, at a specified angle of incidence to the possible ore zone. There are then four differing ways to arrive at the hole path.

1. Directly plot back up hole with simulated survey data based on possible deviation rate. The deviation rate should be varied as the path passes through differing rock types, varying attitude of bedding, and/or foliation, core size, etc.

2. Same as 1 but choose an arbitrary position for a collar on the surface, then plot simulated survey data to arrive at target by trial and error.

3. Overlay the transparent section with target intercept over a file copy of a previously drilled hole in cross section, with similar geology and length. Overlay so that the centre of target and previous ore intercept are coincident, and the target intercept is parallel to the previous hole. Then trace the hole path back to a point of contact with the surface profile, on the new hole plot. Before disturbing the overlay, measure the angle between the horizontal on both plans, at this point on the surface. The declination of the new hole will then be this angle, plus or minus the declination of the old hole at that point.

Remember that the new hole must be drilled with same sizes and tools as the hole which was traced to obtain the same curvature.

4. Work out the entire path on the basis of deviation rate equations from Bluck, 1978, providing one knows the anisotropic strength index. These equations appear fairly horrific, but they really do appear to work, and they are certainly well worth experimenting with. This method is only part graphical, and may well be a precursor to computerised hole path prediction.

Now there should be the planned hole on graph paper. This plot should be annotated with core sizes, bit and barrel types, etc., and maybe desired wedging points.

11.7.2 *Hole management*

1. The annotated planned section can now be overlain with a transparent drafting film. In fact, it is desirable to have a duplicate of this work station to which the driller has access (one at rig and one at field office).

2. As the hole progresses, each survey should be plotted on this transparency, if not immediately, then shortly after. Declination and Azimuth are plotted as being constant over the interval between the midpoints of adjacent survey depths.

For example:

	Depth	Declination/Azimuth
Reading at	surface	60/190°
	30 m	58/189°
	60 m	56/189°
Plot	0-15 m	60/190°
	15-14 m	58/189°
	45-60+ m	56/189°

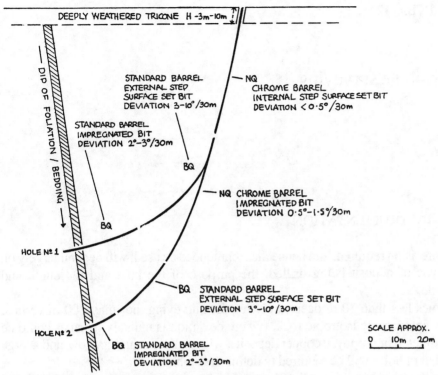

Figure 52. Typical sectional hole plan from Thalanga.

Azimuth will not normally be plotted unless there is significant variation, or a computer is available to assist with calculation.

There are of course many computer programmes which can be used to plot holes but as these mostly require a plotter to show path on a reasonable scale, they are not really useful for the normal early phases of an exploration project, but they would be considered essential in deep mine drilling. One big advantage is that a number of simulated paths can be quickly analysed to compare various options.

3. As the hole progresses, consult with the driller about variations in drilling tools which appear warranted. To test any ideas, the path can be redrawn to see the effect.

4. Annotate the hole plot with actual sizes of core and types of tools used, etc. This will be very useful as programme continues, to assist the planning of further holes (Figure 52).

CHAPTER 12

Downhole surveying

12.1 INSTRUMENTATION

The precision required, and hence the technique to be used, will depend largely on the type of deposit being drilled, the purpose of the hole and its length and attitude.

Holes less than 30 m deep rarely require surveying, nor may 300 m vertical holes, into largely isotropic rock, (where deviation is unlikely) such as could be encountered in porphyry copper deposits, where metal values are low, and a large number of holes will be required to define the orebody.

On the other hand, a small orebody of high unit value metal such as gold, or of high grade, such as a massive sulphide body, will require quite precise data. An error of several metres could significantly alter later calculation of the reserves.

12.1.1 *Acid tube*

A glass tube similar to a culture tube (parallel walled), of soda lime glass rather than Pyrex, is filled with variable strength Hydrofluoric Acid and lowered down the hole to the depth required, and kept stationary for a time. The acid strength used depends on the down hole time needed, and sometimes on the quality of the glass used. The tube is then removed and washed. The acid will have dissolved a small quantity of the soft glass walls, and the meniscus level will be easily seen.

The angle which the meniscus line makes with the tube wall can be measured with a protractor, and using a set of tables, this can be corrected for the variable surface tension effects on the upper and lower walls of the tube (Figure 53) and so the declination of the hole at a point known. These tables are available in books such as the AUST.I.M.M. *Field Geologists Manual*. Such a table will only be accurate for 4% hydrofluoric acid in a standard tube. Cummings (1956, p. 444) provides correction factors for different acid strength and tube types.

This method is slow, messy, not very accurate, and provides no data on azimuth, but it is a very cheap back-up.

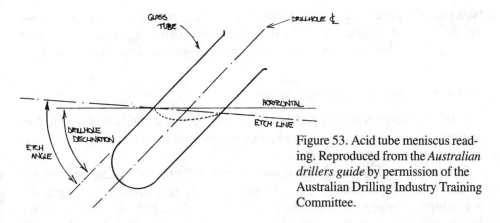

Figure 53. Acid tube meniscus reading. Reproduced from the *Australian drillers guide* by permission of the Australian Drilling Industry Training Committee.

12.1.2 *Tropari type*

This is a compass-clock set in a nest of gimbal mountings. When the clock is wound, the compass-clock is free to move and will come to rest in the horizontal position with compass card orientated. Upon expiry of the time set on the clock, clamps are activated to hold both the compass card and the compass-clock position relative to the gimbal mounting frame. The time taken to activate the clamps can be varied to allow for the time taken to get the instrument to the required depth.

Azimuth and declination are directly read from the instrument using a hand lens.

The major disadvantage of the instrument is that minor dust particles can get into the open bearings and cause 'sticking'. Whilst theoretical accuracy of declination should be $\pm 0.5°$, my experience has been that it is more likely $\pm 3°$. Note also, that as the azimuth is determined by a magnet, then the instrument must be attached to brass or aluminium rod which can extend 3 m or more, beyond the end of the drill rod string and cannot be used inside the casing (except for declination only). Also, there is no permanent record, so that transcription errors can occur. Another common brand of similar instrument is Pajari.

12.1.3 *Eastman type single shot camera*

Although the mechanism is much more sophisticated, this instrument works on a similar principle to the Tropari type, but with the added feature of a photograph of the compass card and clinometer, rather than locking clamps. The clock now activates the light source to allow film exposure.

The small disc film provides a permanent record. Declination is read to $1°$ and can be reasonably estimated to $0.1° \pm 0.1°$. Azimuth is read to $5°$ but can be reasonably estimated to $1° \pm 1°$. These accuracies are more commonly attained as all mobile parts are sealed, as compared with the exposed workings of the Tropari type gimbals. As the azimuth is derived from a magnet, non-magnetic rods must be used as for the Tropari. It will be noted that this is still a single point measuring device. Other common brand names of similar cameras are Humphrey and Pico.

12.1.4 *Eastman type multishot camera*

This takes exposures on roll film at pre-set time intervals, and is thus able to take readings at many depth intervals, with one trip down hole. In all other respects it is the same as the single shot camera.

The latest development in Eastman cameras is the use of an accelerometer to measure declination to an read accuracy of 0.1°. The accelerometer is effectively a gravimeter reading three dimensional field strengths. Data is stored in a data logger which is run with a multishot camera. Upon retrieval the data has to be massaged to produce the required output.

This instrument is not used a great deal on projects with small diameter holes, because the time savings, and hence cost savings, are not so great in relationship to the capital cost of the instrument. Also, probably even more important, is that we need to know downhole survey data, as the hole progresses, so that the hole may be best controlled. So an exception is the surveying of a group of holes such as a group of reverse circulation holes which, although not surveyed while drilling, will require precision if an orebody has been outlined.

12.1.5 *Reflex-Fotobor*

This instrument, produced by the Swedish company, Reflex Instrument *AB*, operates on the principle of measuring the angular difference of light beams from a common source down flexible rods which are reflected from annular reflectors at varying fixed distances from the source. The reflected light (as rings) is recorded on photographic film, which simultaneously records the relationship to the horizontal, by including in the exposure, the level bubble. The instrument is multishot (readings at one or two minute intervals).

After recovery of the probe and development of the film, six angular values are extracted from each frame. Three of these are for dip, and three are for direction. Using a computer or programmable calculator, these are converted directly to three dimensional hole co-ordinates. The fact that it is not direct reading, is one of the disadvantages with this method, as is the absence of a single shot model, with less capital cost. The method requires precise initial azimuth and declination of the probe to be determined by normal survey methods so that the ensuing data will not compound any errors. The data is recorded as angular changes like a normal survey traverse.

The major advantage of the instrument is that the direction is measured non-magnetically, which means it can be used inside casing, or rods, and in magnetite and pyrrhotite bearing orebodies.

12.1.6 *Reflex Maxibor*

This is a development of the Fotobor in which, instead of a camera, there is a *CCD* sensor (miniaturised video sensor) of the reflected light rings. This can record 16384 pixels on the one square centimetre sensor. An on board computer measures the signals and compares with a liquid level sensor. Up to 4000 readings can be stored in the 256K of RAM. This virtually means continuous readings but it is usually read at 3 m intervals for the 6 m instrument (3 m between rings) and 1.5 m for a 3 m instrument.

Upon retrieval the memory is downloaded onto a laptop site computer which, within minutes will provide path parameters for every 3 m (or 1.5 m) downhole or a plot can be made directly.

This method like the Fotobor, is similar to a surveyor making a traverse, i.e. transferring direction and declination by foresight and backsight in an interconnected chain of readings. For this reason the position of the end of the hole is only as accurate as the original azimuth and declination entered as the start readings. Note it is the attitude of the instrument at the start which is important and not necessarily that of the attitude of the collar.

12.1.7 *Gyroscopic compasses*

Devices similar to the Eastman type multishot camera are available with gyroscopic compasses. These have only recently been made for the smaller hole diameters, used in the mineral exploration industry. Holes down to *NQ* (inside rods) are now accessible. They are lowered downhole on a wireline power cable and take periodic photographs like a multishot Eastman. An accuracy of $\pm 0.25°$ in inclination and $\pm 2.0°$ in azimuth is claimed. It should be noted that they cannot be used for declinations of less than 45° as bearings will wear unduly.

As these are available in Australia only with operator, it is quite costly to do hole by hole, or to prepare holes with slim PVC casing for some later bulk measuring effort. These then will only be used on a 'have to' basis for deeper holes in magnetic host rocks.

12.1.8 *Miscellaneous devices*

These are all workable devices which have been outdated, but it is useful to be aware of the principles, should there be problems with existing equipment.

Carlson compass
This is a gimbal mounted compass-clinometer very much like a Tropari. It is fitted into a tube with gelatine or agar which, after a residence time of 1.5 to 2 hours, sets and holds the gimbals steady, so it can be withdrawn and read. However, do not do this with a Tropari, as the clock will not appreciate being immersed in the liquid.

Apart from the long time and mess involved, gelatine will not set at elevated temperatures.

Maas compass
A compass ball is floated in gelatine or agar, in a partly filled culture tube. The declination is obtained from an acid tube, in a two chambered tube.

Radiolite compass
These were very similar to a single shot Eastman. In these, there were no bulbs for light, but radioactive paint on compass and clinometer needle. These gave very fuzzy pictures.

Carlstrom-Bergstrom instrument
This, like the fotobor, was invented to avoid the use of a magnet. It consists of a long, somewhat flexible tube which has minimum hole clearance, in the centre of which a thin wire is tensioned. The relationship between the wire and tube wall, as the rod bent to conform to the hole, is able to be recorded on film, and deviation calculated.

Kiruna method
This electroplates copper on to a tube wall in a similar arrangement to an acid tube, but it requires a little current, and therefore a cable.

Alignment clamps
An acid tube is loaded, and its orientation is noted before lowering. As the rods are now lowered, the azimuth is maintained using special clamps which effectively are a parallel telescope along the rods. This is an alternative for magnetic terrains – developed by the US Bur. Mines (Report of Investigations No.3773).

Eastman type W mechanical drift indicator
This makes a dot on a special paper disk by means of a plumb-bob, incorporating a depressible stylus. This is good for high temperature ground, where photographic methods will not work.

Eastman type M photographic drift indicator
These were made to suit near vertical holes. They work on the principle of a light source imaging a plumb-bob on a photographic disk.

Sonar telemetry
This is still in a development stage, and whilst details are still hard to get, it is expected that this will enable continuous surface readout of survey data. There will be no electrical connection between the surface and down hole part of the device, which itself will not preclude coring at the same time.

12.2 CONDUCTING SURVEYS

It is important in most cases that downhole surveys be done as the hole proceeds. The principal reasons for this are:
- Should unplanned deviation occur, then corrective action can be taken;
- Casing may be inserted, and make azimuth reading impossible, unless a Reflex-Fotobor/Maxibor or similar is available.

Surveys should be taken every 30 m, more or less, depending on the deflection rates either planned or being experienced. Another guide is to have sufficient surveys, to reduce the difference between adjacent readings to less than 2 °. It is not critical that each survey be at precise multiples of 30 m, the survey depth can often be one which suits the drillers. For example, if there are 1.5 m from collar to top of head, and there is a 3 m barrel plus 1 m of reamers etc., and rods are three m long, then the most convenient depths for the drillers will be 3n + 3 + 1 + 1.5 (n = number of rods). So surveys will be at 29.5 or 32.5 m, and then 59.5 or 62.5 m, and so on.

For acid tube, and single shot camera surveys and the like, it is best to leave these to the drillers to do as they progress, and that checks be made at random intervals to audit the results. With the Tropari and similar instruments, in which the instrument cannot be reused without resetting and destroying the physical record of the result, it is best to visit the rig at frequent enough intervals to read the instrument personally on each occasion, and have drillers check the reading.

Maxibor, gyroscope or multishot surveys are best done at the end of the hole preferably with rig on site. If there are severe deviations planned or being experienced in very magnetic ground then they will have to be done from time to time as the hole progresses so that the hole control decisions can be made with realistic data.

At regular intervals (possibly weekly), check any instrument and its accessories for thread, and o-ring water tightness. Ensure it is kept in a safe place when not in use. Ensure spare batteries, film, developer, acid, etc. are available.

Check the accuracy of any method and/or instrument used. Azimuth should be checked in each of the cardinal directions, at least, so that a correction table may be organised. These checks should be repeated if any difference in the instrument configuration occurs, e.g. a change of battery type or brand, or component exchange. It should be noted that, commonly, the error will be variable in different directions, just as we see noted on the face of aircraft compasses. Figure 54a shows a typical error distribution in different directions at one mine.

To enable quick and accurate checks to be made, a checking station or device should be established. An outline of such is as follows (Figure 54b). Firstly select a location well away from any cultural effects, such as buildings, fences, powerlines, etc. and an area with no magnetic disturbance from laterite in soils, or any other reason for magnetic 'noise'. Then, erect a post of nonmagnetic material, about 800 mm out of the ground with a central hole to fit an aluminium rowlock,

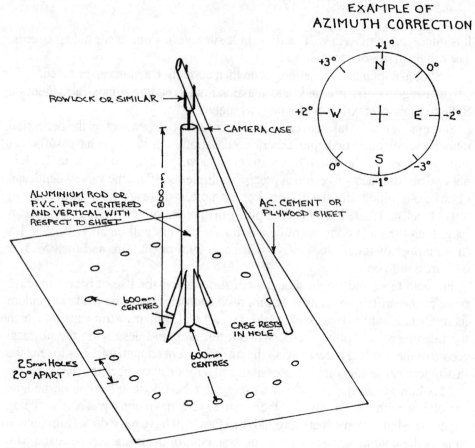

Figure 54. Checking accuracy of survey device.

or similar device, on which the survey instrument rod can be supported, but can rotate about a central axis. A number of holes are drilled in a plywood sheet which has a central hole to fit over the post. There are holes positioned on a circle at about 600 mm radius from the centre of the large post. These holes allow a recessed block to be placed to allow the toe of the instrument carrier tube to rest in a fixed position, which ideally would be the cardinal compass points.

The plywood sheet must be accurately aligned on a known bearing. This is best done by having pegs about 20 m apart placed exactly on the required bearing and then a stringline between the two passed over the sheet.

For instruments such as Maxibor it is recommended that the following method of starting point alignment be used. The example is for an *NQ* rod string and can be appropriately varied for rods of a different size.

1. Rest rods on bottom and release from drill, making sure that the rods do protrude past the casing. 1 m is ideal to stop rod droop downhole at the start.

2. Place a pair of *BQ* (or even *NQ* would do) rods under the end of the protruding rod string and resting on the collar of the casing. The other end of this pair can rest on the mast or rod slide in a couple of places. This will provide an extension of the rod string with a continuity of azimuth and declination(does not have to be precise). This pair of rods requires clamping so that once surveying commences there is no further movement.

3. Insert maxibor in its instrument case (about 7 m long) so that about 4.5 m are out of the hole and lying snugly in the groove formed by the pair of rods discussed above. This will allow the surveyors to sight the top and middle reading ring positions (3 m apart) still above the casing. To find the centre of the Maxibor case use micrometer callipers so that the diameter can be determined and then sighting can be done on the half way point.

4. The distance from the survey instrument usually some 10 to 20 m in front of the rig can be precisely measured to the above two points on the instrument case. The best is to use a dismantled *EDM* reflector with a cone shaped rear to position exactly on the points to be measured.

For a gyrocompass and others which also need a preset starting direction and azimuth it is not necessary that the original attitude has any relationship to the drillhole path, only to an initial instrument reference reading. It is therefore best measured by the surveyors almost completely out of the hole at a dip just a little more than 45°, using a similar method as described for the Maxibor in paragraphs 3 and 4 above.

CHAPTER 13

Orientated coring

This means the cutting of core so that when removed, it can be orientated to the in situ dip and strike of contained structures, etc. There are four basic techniques being used. The selection of method depends on the type of rock being cored and the accuracy required.
- Residual magnetism;
- Drill rod alignment;
- Core marking;
- Orientating core barrels.

Residual magnetism
Rocks with a strong residual magnetism can be realigned by measuring this, providing the characteristics of source formation, or pluton, are known.

Drill rod alignment
This involves carefully breaking off the core in a conventional single tube barrel. The rods are then carefully withdrawn, such that the rods are not rotated, and maintain their orientation by means of clamps, and a parallel sighting device (see Figure 55). This is very time consuming, and there is plenty of scope for error. The same system can be used to approximately orientate a wedge by alignment of rods into the hole.

Core marking
These methods involve the marking of the hole face on the low side of an inclined hole before it is cored. Variations available are:
1. *Punch marking.* A punch or chisel is lowered down the hole and dropped so that it makes a mark on the lower side of the hole face. The punch is usually about 1 m long and 25 mm thick with tapered end and a hard point. The idea is not to mark the very bottom of the hole (which is worn away by the bit), but just above so that the mark will be preserved in the core (See Figure 56). This is quite a good method, but not very accurate, and not good for either very hard rocks (difficult to mark), or for very soft rocks which break.

106

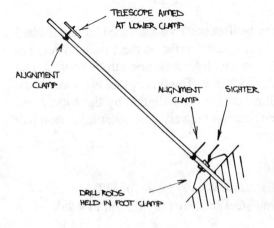

TELESCOPE AIMED
AT LOWER CLAMP

ALIGNMENT
CLAMP

ALIGNMENT
CLAMP

SIGHTER

DRILL RODS
HELD IN FOOT CLAMP

Figure 55. Alignment clamps. Reproduced from the *Australian drillers guide* by permission of the Australian Drilling Industry Training Committee.

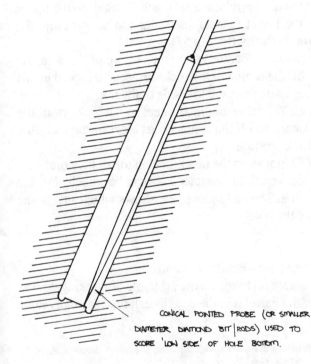

CONICAL POINTED PROBE (OR SMALLER
DIAMETER DIAMOND BIT|RODS) USED TO
SCORE 'LOW SIDE' OF HOLE BOTTOM.

Figure 56. Core marking punch.

2. *Pencil marking.* Uses the same tapered bar, but the steel tip is replaced by a holder in which a crayon or chinagraph pencil is held. The colour can be varied to contrast with the rock colour. This will work well on very hard core, but it too, will have some difficulty with soft core.

3. *Dye or resin marking.* Much the same as previous two methods, but uses a somewhat different type of rod now to accommodate a syringe type device to squeeze out dye, or two part fast setting resin. This is much more messy, but it is the alternative suitable for soft friable core.

Scratch survey

This requires the bottom of the hole to be flattened by a flat faced bit and flushed clean. A survey instrument with a non-magnetic scribe on the base is lowered to the bottom. After a wait for a survey reading, hydraulic pressure is applied, and the scribe makes a radial scratch on the rock face. The survey tool is removed and normal coring continued. The resulting core can be aligned by the survey and orientated. Such a method seems quite accurate but very time consuming, and will only work in reasonably competent rock.

'Craelius Core Orientor'

The following description is from *Australian drillers guide* (p. 372).

'This relies on a small free-moving steel ball, identifying the low side of an angled hole.

1. The *CCO* is inserted in the core barrel inner tube and 'locked' to the upper edge of the core lifter. When the barrel is assembled, the contact pins and the pressure rods will protrude through the bit (Figure 57).

2. As the barrel is lowered (without rotation) on to the bottom of the hole, the contact rods are positioned to 'fit' the profile of the rock surface and locked in this position; the orientation marking ball marks the low side of the hole.

3. Applying weight to the *CCO* causes the instrument to 'unlock' from the core lifter. The pressure rod extends to lift the contact rods clear of the rock (the rods retain their record of the rock profile).

4. As the core is cut, the *CCO* moves up the inner tube on top of the core.

5. When the inner-tube is recovered, the contact rods can be matched to the profile of the upper end of the core. The ball indentation shows which side of the core comes from the low side of the hole'.

Orientating core barrels

There are two principal variations here – continuous and batch.

Continuous orientating barrels are all based on the principle of marking core as it passes through the bottom of the barrel (i.e. the core lifter); this is usually done

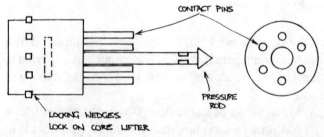

Figure 57. Craleus core orientor. Reproduced from the *Australian drillers guide* by permission of the Australian Drilling Industry Training Committee.

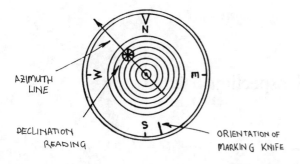

Figure 58. Camera card showing orientation mark. Reproduced from the *Australian drillers guide* by permission of the Australian Drilling Industry Training Committee.

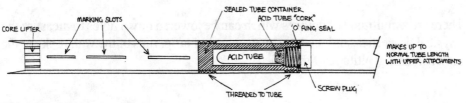

Figure 59. Batch orientating innertube.

with a knife. This means that all the core in a run can be orientated with respect to each other. A survey device such as a down hole camera or acid tube can relate this knife mark to the azimuth (Figure 58).

Batch orientating barrels are based on holding the core in the lifter, and placing the orientation mark on it when it reaches the surface. The most common type now being used has a three part inner-tube (Figure 59). The lower section has a core lifter, like any other inner tube, and can have thin slots on one side, which are lined up with a mark at the base and a mark at the top. The mid section contains a chamber for an acid tube. This chamber has a keyway aligned with the top mark referred to already (and hence aligned with the slots). The acid tube used has a key (piece of wire glued length-ways along the outside) so that it can only fit into the keyway one way i.e. the acid tube can be aligned with the slots in the lower tube. The upper tube is only to make up normal length to enable retrieval.

The inner tube is lowered and cores about 300 mm, then the barrel is carefully pulled to break core, but not to rotate. The barrel is held for 15 minutes or so, to allow the acid to etch a meniscus mark. The inner tube is then removed in the usual way, and the acid tube quickly removed and washed. A hacksaw blade or knife is then put into the slots and the core scratched with a mark which is now aligned with the key on the test-tube. From the meniscus we are now able to relate the mark on the core to upper or lower side of an inclined hole. A set of annular protractors, to fit core and test-tube, makes an excellent method of transferring the necessary rotational angles to calculate the dip and strike of any planar feature.

CHAPTER 14

Downhole logging and inspection

There are a multitude of devices which can be lowered down holes to measure the physical properties, and recently even chemical properties of the wall rocks or the dimensions and attitude of the hole.

14.1 DOWNHOLE LOGGING

It is not within the scope of this book to describe the theory and uses of such logs, but here are some of the features directly related to the drilling of the hole.

Magnetics – only inhibiting factor is steel casing.

Gamma Ray – no inhibiting factors.

Self Potential – affected by fluid salinity and must have some liquid in hole to transmit currents. If used for aquifer detection then works best with saline fluids. Casing must be absent or slotted.

Resistivity – affected by muds, so only get apparent resistivity. Casing must be absent or slotted, to allow contact with wall rocks.

Electromagnetic – steel casing must be absent – PVC can be used.

Neutron – no inhibiting factors.

Temperature – hole with fluid of uniform composition. Must await end-point of exothermic reaction such as cement setting (wait 12 hours).

Calliper – used to log internal rod or hole diameters. Mostly seen in water and oil holes. Figure 60 illustrates the general principle. As long as fluids in hole are not too viscous they have no effect.

14.2 DOWNHOLE INSPECTIONS

There are many circumstances where one wishes to know something about what is down there. There are a number of devices to do various specific tasks, and I am sure that ingenuity will provide more as needed, but here are a few.

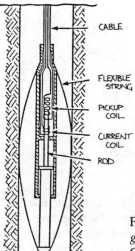

Figure 60. Hole callipers. Reproduced from the *Australian drillers guide* by permission of the Australian Drilling Industry Training Committee.

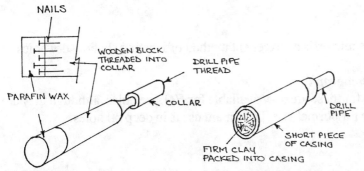

Figure 61. Impression blocks. Reproduced from the *Australian drillers guide* by permission of the Australian Drilling Industry Training Committee.

Impression blocks
Two species of these are illustrated in Figure 61. These are mainly used to get impressions of objects stuck in the hole (wall or bottom).

Wall impression tool
This is illustrated in Figure 62 and is used to get impressions of wall irregularities on a thermoplastic film over a layer of resilient foam. This is good for locating joints, bedding, faults, veins, and resolving core loss problems.

Periscope
Shallow water bores and site investigation holes can make use of these. They have to have a downhole light source, usually batteries.

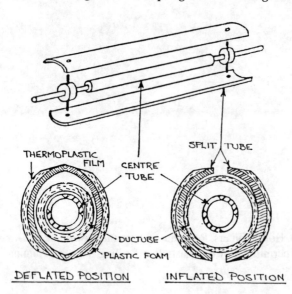

THERMOPLASTIC FILM
SPLIT TUBE
CENTRE TUBE
DUCTUBE
PLASTIC FOAM

DEFLATED POSITION   INFLATED POSITION

Figure 62. Impression packer. Reproduced from the *Australian drillers guide* by permission of the Australian Drilling Industry Training Committee.

Mirror

Sun, reflected by a mirror, is an excellent method of viewing shallow dry holes.

Television and Photography

Downhole probes for these are now available for *HQ* size holes, although they are still too expensive for normal use. Their main use is in deep oil holes.

CHAPTER 15

Sampling

15.1 ROTARY AND HAMMER

These methods produce rock chips, or clay shavings, and similar disturbed material. These may arrive at the surface dry – with air (± mist) or wet – with water (± mud, foam, etc.).

The chip size is controlled by:
- Rock type;
- Rotation speed;
- Penetration rate;
- Type of bit;
- Hammer type.

Generally, larger chips are more desirable as they use less energy per given weight, to cut, and they make geological logging easier. Conversely, larger chips are more difficult to clear from the hole. There is, therefore, always a compromise.

Figure 63 illustrates how chip size is related to rotation speed and thrust (penetration rate), the two easily controlled variables. The rock type in this example is not known, but it does show the effects.

For roller bits, Figure 64 gives some idea of the range of rotation speeds and thrusts applicable in various formations. Greater thrust gives bigger chips.

For hammer drills, the best way to get larger chips is to have a sharper bit, although rotation speed may change the size a little, as will the thrust. I have noticed that face sampling reverse circulation hammers give bigger chips.

15.1.1 *Sampling the chips*

Once the chips arrive at the surface, the sampling process depends on the accuracy required, which relates to the purpose of the sampling (e.g. geochemistry, ore reserve assay, etc.), and the value or type of material (e.g. high unit value such as gold, medium such as copper, or low such as roadmetal).

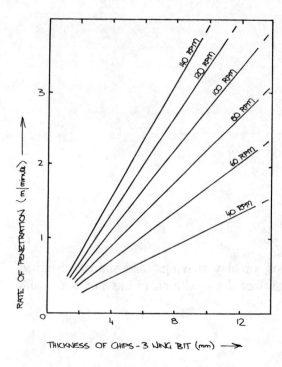

Figure 63. Chip size variation with blade bit. Reproduced from the *Australian drillers guide* by permission of the Australian Drilling Industry Training Committee.

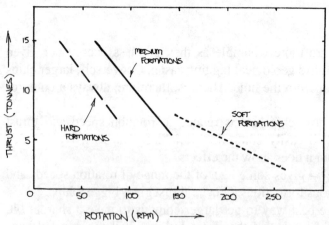

Figure 64. Roller bit thrust for various rocktypes. Reproduced from the *Australian drillers guide* by permission of the Australian Drilling Industry Training Committee.

There are, of course, infinite combinations and variations, so a few typical situations will be discussed, which will illustrate the concepts of good sampling.

RAB reconnaissance geochemistry for Cu, Pb, Zn.
Rotary Air Blast, with blade bit, to drill through 30 m tertiary alluvials to bedrock, which may contain massive Cu, Pb, Zn.

As we will be sampling each 3 m interval for the entire hole (palaeodrainage sampling), and the requirement of geochemistry of medium unit value metals, we do not need precise samples, neither in terms of depth (no need to clear hole at each sample interval) or representivity.

Collect sample at collar in small bucket, dishes, or on shovels, from unrestricted natural fall-out from hole, but try to ensure that the receptacle is positioned so that:

1. it gets filled gradually over the entire sample interval, and

2. it contains a fair proportion of fine and coarse e.g. a collection tray should have its long axis in a radial position from the hole.

A grab sample, of say, four handfuls of such material, could be adequate for assay, with a small sample of coarse chips being kept for geological record.

The residue can be dumped on the ground in heaps aligned in groups of 10 samples or so. This material can provide back-up samples for a few days. After rain washes such heaps it is sometimes convenient to relog as a larger quantity of clean chips are visible.

RAB reconnaissance geochemistry for gold
Reverse circulation air core through overburden and saprolite, and convert to downhole hammer in hard rock.

Because we are dealing with a high unit value material, and as in this situation there may be a nugget effect, we must be more careful.

The hole must be fitted with a collar, and chips will pass to a cyclone to separate solids to bottom, and air (with some fines) to top. We would now catch all the sample (cyclone underpass only), from each metre in a large plastic bag. We need not be too concerned with hole clearing at each sample interval, as we are only looking for order of magnitude of assays.

Because of the high cost of gold analyses (compared with drilling cost) there are then two common paths for reconnaissance drilling.

1. Make a composite of each four consecutive intervals. This can be done using a 50 mm PVC tube inserted into the large sample and so taking a core of equal length from each, and blending these into one sample. Should the initial sample results show any significant values, then the relevant 4 m inverval can be resampled using a Jones Splitter to obtain a more accurate split of each single metre sample.

2. Sample taken from a longer interval – 3 m is common (rod length multiples). In this case it is more likely that a Jones Splitter would be used to obtain a 2 kg to 5 kg sample for assay.

Again, larger chips from each interval (before any compositing) should be sieved out, logged and retained for geological records. The residue is usually retained in the collection bag for several weeks, either on site or in core yard. Remember, cattle can cause havoc, and sun can cause bags to disintegrate if exposed too long.

Hammer drill for ore reserve quality samples – Cu, Pb, Zn
In this situation, accuracy is much more important, as the results form the basis of investment decisions, and following this, day to day mine planning and production decisions.

As before, the actual drilling and sampling techniques used will depend on the grade and size of intersection expected, as well as the uniformity of these parameters from place to place within the deposit. We will now look at three examples from both ends of the spectrum.

High grade base metal mine (e.g. Thalanga, Aust.). Ore grade ranges approximately 4% to 20% Cu equivalent and ore lenses range from 2 to 20 m wide. Drilling of near surface openpitable ore was done with triple tube *HQ* core, and infill with downhole hammer reverse circulation, with in-pit drilling by open hole downhole hammer. The core was sawed, and half sent to laboratory (1 m interval). *RC* drilling samples were collected by cyclone (1 m intervals), and split with a Jones Splitter to 2 kg sample, the pit drilling was sampled by cyclone collection of coarse fraction (2.5 m intervals – 5 m holes) and then crude cone and quartering to a 2 kg sample. Tests were done to confirm that the absence of fines and the crude splitting were adequate.

Note that wet percussion samples, both open hole and *RC*, gave unreliable results, and holes in dry cavernous leached gossan gave very poor recovery, and thus were of doubtful value.

Large low grade base metal mine (e.g. Marcopper, Philippines). This is a low grade porphyry copper mine with grade of the order of 0.7% Cu.

Inpit drilling (1970) by open hole downhole hammer, with no collar or cyclone, and sample collected by grab sample from the ground at collar (approximately 8 kg coned and quartered to 2 kg.) Sample interval was 3 m with 9 m holes.

Medium grade gold mine (e.g. Mount Leyshon, Aust.). This is a stockwork and disseminated gold in pyrite deposit, which is mining mainly unoxidised ore (1990).

Production drilling and sampling is by open hole downhole hammer, 150 mm diameter, 10 m deep, sampled at 5 m intervals. Sample collection is done with radially aligned dish under unrestricted fall-out.

15.2 DIAMOND CORE

In general, of course, the objective is to have a cylindrical 'core' cut from the rock through which the drill passes.

The equipment in which this cylinder of rock is collected and brought to the surface is called a core barrel. The drilling is essentially a batch process in which the hole advance continues until the barrel is full, or because of blockage, etc., and will not take any more. At this time advance stops, and the core is brought to the surface.

In the original drill rod strings this was done, and in some special situations, still is done, by bringing the entire rod string to the surface to get at the core barrel. Modern wireline systems allow the core to be retrieved by pulling only the innertube through the rods. This is done by lowering an overshot on a wire, which latches on to the upper end of the innertube.

The very simplest core barrel then, is a single tube barrel, and the complete assembly consists of a bit, reamer, corelifter, and barrel. The reamer provides two functions, one, to maintain the correct hole diameter to ensure rod clearance, and two, to provide seating for the corelifter. The corelifter consists of a tapered spring-like device, which allows the core to travel upward through it, but when core attempts to go down, it wedges in and clamps it. It then not only prevents cut core from dropping out, but it also allows the core to be broken off from the bottom.

From this very simple device, we move to a double tube barrel, which diverts the drilling fluid to the bit by way of the annulus between the barrel and innertube. These were devised for broken or friable ground, where core could easily be washed away.

In later models the innertube was fixed with a ballrace so that it would not rotate with the rods, and so cause grinding and wearing of core. Then followed a whole collection of barrels which refined the idea of core not rotating and entering the innertube as soon as possible to avoid washing. The corelifter was now transferred to the innertube as part of this concept. A split innertube became available to allow the better transfer of undisturbed core into trays etc. These barrels were identified by *M, L, ML, MLC* appended to the rod size (e.g. *BMLC*).

The wireline barrels now in use are an adaptation from these double tube core barrels (Figure 65). The innertube has a spear head which allows the overshot (Figure 66) to attach itself to the innertube to retrieve it. Within the innertube is a cut off valve which exaggerates the effect of a blockage with very high water pressure on the gauge, alerting the driller to the problem.

For that undisturbed core it is now necessary to use a triple tube barrel with split innertube. This split innertube is usually removed from the innertube by hydraulic pressure from the rig water pump. Triple tube systems are identified by appending to the rod size (e.g. *NQ-3*). The additional tube means a slight reduction in core diameter must be made.

Core barrels come in a variety of lengths. 3 m (contained core) is most common, but for deep holes 6 m barrels are often employed, and for short holes with large diameter, barrels down to 1 m may be used for ease of handling. Short barrels may also be used to obtain a shorter active length in the rod string to obtain a greater deflection rate.

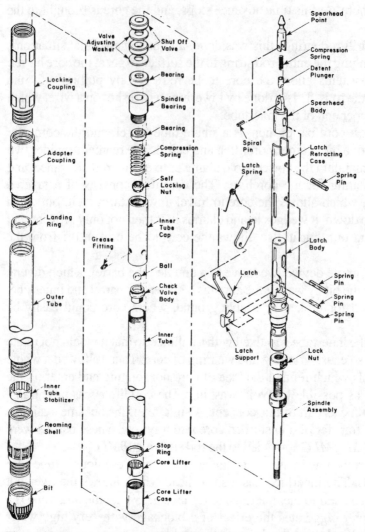

Figure 65. Diamond core barrel nomenclature. By permission of Longyear Australia.

15.2.1 *Core size*

Core size will depend on the bit and barrel combination in use. The more common sizes can be seen in Table 1.

The size and quality of core to be obtained will depend on the following parameters, which must be considered before a hole starts as it is very costly to ream out a small diameter hole to get larger core:

What is the purpose of the core?
For example, normal *NQ* core may be adequate for assay and geological data on a

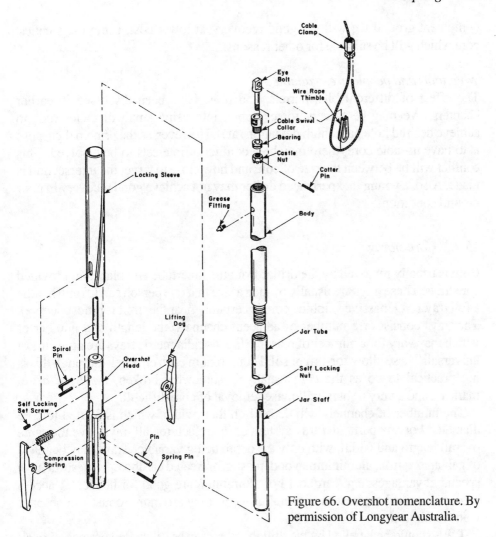

Figure 66. Overshot nomenclature. By permission of Longyear Australia.

certain project. But for geotechnical purposes it will be necessary to have undisturbed triple tube core, so either use *NQ3* or *HQ3*. On the same project we may need larger pieces of core for metallurgical work and so may find *PQ* necessary.

What cost constraints are applicable?
If we ignore hole paths or depth, then the larger the core diameter, the more useful is the core (better recovery, less disturbed, larger more accurate sample, etc.). The cost of core is approximately proportional to the diameter, and this must be measured against the budget.

What quality of rock is to be cored?
In broken ground, recovery will be better with the larger core size. In very hard

competent ground we will get good recovery, at lower cost, using the smallest core which will be suitable for other reasons.

What tools and people are available?

The effect of different bits, barrels, and rods, have been discussed in earlier chapters. We may be forced, for example, into using small diameter rods to achieve desired hole path (high deviation rate). However, as the principal purpose is to have useable core, there may be a conflict of interests to be resolved. This conflict will be between the size of core and how close to target the intersection is made. Also, a young inexperienced driller may not achieve good recovery in poor ground conditions.

15.2.2 Core storage

Core is usually removed by the driller from the innertube and placed in provided core trays. These trays are usually of such a size that one person can lift and handle a full tray. (*HQ* massive sulphide core is certainly near the limit for most people). The trays consist of a number of adjacent channels, the length and number of which may vary for a number of reasons. In Australia metric trays are now almost universal. These allow for 1 m of solid core to fit in each channel, although this is not practical to do as the core usually is somewhat broken, there are depth markers, and a very tight fit will make removal of core difficult.

The number of channels will depend on their width, which is related to core diameter. For any particular tray system or manufacturer, all trays have the same overall length and width, with only the depth being variable. Trays may be made of galvanised iron, aluminium, wood, plastic, or treated cardboard. These all have special advantages, e.g. cardboard or aluminium are good for helicopter access sites, galvanised are best for general usage as they are more robust and cheaper than aluminium.

Trays should be labelled by the drillers, (they can be relabelled permanently by project staff within a few weeks) preferably on the left hand end of the tray. Here should appear the hole number, and either the tray number, or the 'from and to' metreages. The core should be placed in the tray as a book is read, which is to say the metreages should increase across the tray left to right and down the tray. Remember, the core being removed from the innertube is put in 'backwards', as the core drilled last is normally emptied first.

Depth markers or coreblocks should be placed by the driller after each run is placed in the tray. It is important that all individual runs are so marked that any non-recovered portions, such as those due to cavities, can be accurately known.

It is not uncommon for drillers who have been making regular 3 m runs, to not place markers when they encounter a zone of broken core with many short runs. I have heard a geologist say that there is no need to put in all the markers, as it just consumes more tray space. These practices are not to be tolerated; all runs must be marked. The markers should have the metreage to two decimals, and be done in a manner that they can be read several years later.

The markers should be nearly the channel size in section, and should be of such a size and weight that they will not easily blow away or jump about in transport. So do not use paper, cardboard, moulded or other plastic; do not use small pieces of timber (e.g. cut grid pegs), green timber (sawn local branches), rough sawn timber, round blocks in square section and *vice versa*. I prefer cured hardwood block dressed on two sides, one to be written on with felt pen, whilst the other has an aluminium tag to indent write with a biro or nail. I also suggest that the metreage be written in the bottom of the tray, so that, if the block does get removed, it can be easily replaced.

So now the core is ready for delivery, or handover to the client, normally the core grabber or geologist. Because long or rough transport may disturb the core, there may be some logging type activities which are best done at the drill site if the core is soft or broken. Such items are core photography, logging Rock Quality Determination (*RQD*) and other structural and geotechnical aspects.

Core trays which have to travel any considerable distance or over rough terrain such as in a mine underground, will require lids attached.

So let's take the normal case where the core is brought back to the core yard or base camp for processing.

15.2.3 *Core preparation*

Firstly, if any core orientation has been done, then these pieces will require study to determine which side of the core is up (i.e. facing the surface). Then, by fitting adjacent core up and down hole as far as good breaks allow, as much core as possible is orientated. If no orientated core was drilled, the core should still be fitted and rotated to a common, but unknown orientation. If there is an identifiable foliation or bedding, then it is often helpful to arrange the core so that this planar feature is orientated in a similar attitude throughout (such that the poles to the plane are at right-angles to the camera axis).

Of course the drillers should have the core cleaned, but it will often require a further clean down, as the fitting and rotating will show uncleaned surfaces. So now that it is all nicely joined together and clean, and in a short time, dry, the core can be measured and metreage marks written on core with felt pen, or on some temporary marks such as flagging tape. In the process of doing this, the core recovery can be noted specifically related to each run, but also estimates are made if possible, exactly where any loss may have occurred, such as at a shear, or where there is core grinding.

15.2.4 *Core photography*

Now the core is clean, orientated and depth marked, it can be photographed.

Photographs or slides are most useful for those doing engineering studies, to have a record of core as undisturbed as possible, showing breaks, fissility, etc. To that end primarily, core should be photographed as soon as possible, and certainly before any sampling. A high quality, preferably automatic camera, is recommended. Floodlighting can be useful, but light shade adjacent to bright indirect sunlight will give quite good results. Wet core is usually best, but check it out yourself. It is important to run a few check photographs when involved in a new project, as there may be strong contrast in reflectivity between shiny aluminium trays and dark core, for example.

So that the camera is at right angles to the core (in trays), and at a constant focal length, a frame should be constructed to hold both core and camera. The core tray should be clearly labelled (Hole No. & Metreage). The camera should have a remote cable attached, to avoid camera movement, when operating the shutter.

15.2.5 *Core logging*

Logging is the describing of the character of the rocks which make up the core. This may be done in writing, on an audio tape, or keying the data into a computer disc, or, in fact, it may be a combination of these. Each organisation has its own system which allegedly suits its purpose, but in general, most are logging in codes suited to computers, and using custom made or modified software. Because of the wide variations being used, it is not realistic to describe in any detail how it should be done.

Remember we are logging facts, and where interpretation is given, it should be so noted. An example would be that 2 m of dolerite with parallel margins within a 100 m interval of granite should not be called a 'dolerite dyke' unless there is almost absolute supporting evidence such as chilled margins. It could have been a xenolith or a fault block without such evidence. It should be logged as 'dolerite' – 'possible dyke'. A computer logging system must be able to accommodate these subtle features.

At Hellyer Mine, Tasmania, Australia logging is by code on to forms (1:200 scale) which are later manually entered into computer files (Downs, 1990). Mount Isa Mine, Queensland, Australia (Bartrop & Sims, 1990) uses a similar technique. Pasminco Mining, Broken Hill, New South Wales, Australia, use laptop computers to directly enter coded data (Cook, 1990).

Core is best logged in sunlight, and if this is just too uncomfortable, then in light shade adjacent to sunlight. The core should always be logged whilst wet. It is often useful to relog several times to ensure that those comparative features are all logged or coded in the same perspective. For example, we want to ensure that what was called very siliceous on Monday is more or less the same in core logged on Friday.

15.2.6 *Core sampling*

From the logging, those core intervals which have any possible value are usually known, and in fact, grade estimates should have been made for all intervals of similar grade. Depending on the size of the ore deposit, the order of magnitude of grade and reliability of visual estimates, there will be a great variety of 'correct' ways to select assay intervals, which will also vary with the experience of those managing the project.

Normally we should consider 1 m basic sample intervals, although in large low grade open cut mines such as Mt Leyshon, a 2 m interval is used. These basic intervals should only be divided further if there are visually estimatable significantly differing grades. These extra samples are particularly important at the edge of the orebody. (At Mt. Leyshon, Queensland, Australia, where visual estimates are of doubtful accuracy, they sample the probable margins at 1 m intervals).

Let us look at a hypothetical log (Table 14) of a hole which is to have sampling at 1 m intervals and within which visual estimation of grade is easy e.g. base metal. Note how extra samples are filled in to accommodate differing visual estimates.

So now that the assay interval has been selected, we have to decide how much of the core will be sent for assay. There are a number of factors to be considered, and each project, or even each stage of a project, will require a different approach.

There are two broad but conflicting axioms which we need to consider.

1. The larger the sample, the more representative the assay will be (i.e. more accurate).

2. To enable repeat sampling as checks or for metallurgical purposes, or relogging for some special purpose such as geotechnical, it is desirable to retain as much core as possible, with as little disturbance as possible.

The most common solution, particularly in the early stages of a project, is to split the core along the long axis with a diamond saw or in some cases a stone chisel. Half is sent for assay and half retained. Should check or metallurgical sampling then be necessary, the retained half can be resplit into quarters.

There are situations where quarter core is adequate for normal assay, and if there is some urgency to have greater weight of metallurgical sample, in this case send half core for metallurgy, quarter core to laboratory, and retain quarter.

When drilling porphyry copper deposits in Papua New Guinea it was necessary to send whole core for assay. This was because the mineralization was present as smears on numerous joints with good parting. This meant that in the normal cutting process mineralization was washed away and also remained in the bottom of the trays, so that assays could be reporting about 10% lower values (the profit margin). A small skeleton core collection was kept of pieces of whole core of each rocktype. This would have been no more than 2% of the core in pieces 5 to 10 cm long.

Table 14. Simplified assay log example.

Metreage		Visual estimate of grade	Assay interval (m)	
from	to		from	to
0	96.5	Barren	93	94
			94	95
			95	96
			96	96.5
96.5	104.2	High grade	96.5	97
			97	98
			98	99
			99	100
			100	101
			101	102
			102	103
			103	104
			104	104.2
104.2	108	Medium grade	104.2	105
			105	106
			106	107
			107	108
108	114.4	High grade	108	109
			109	110
			110	111
			111	112
			112	113
			113	114
			114	114.4
114.4	118.3	Medium grade	114.4	115
			115	116
			116	117
			117	118
			118	118.3
118.3	123	Low grade	118.3	119
			119	120
			120	121
			121	122
			122	123
123	150	Barren	123	124
			124	125

For high unit value deposits it is also common to use whole core. It was a practice at Witwatersrand in South Africa to drill N core through the ore, then repeat the intersection with two closely spaced wedged holes and then use two whole and one half cores for assay and retain only one half core.

Sometimes it is necessary to take very small samples for geochemistry, or as a check for that mineralization which is not readily apparent visually. It is common, in these cases, to cut either a small channel with the core saw and collect the cuttings as a sample, or to cut a sliver of core off one side. The main reasons for these techniques are the reduction in core cutting and sample preparation costs.

Nuclear gammaray resonance scattering devices (bulk analyser) now being used by Mount Isa Mine (Queensland, Australia) for copper, and at Renison Tin Mine (Tasmania, Australia) for tin, enable very cheap, accurate, nondestructive core analysis (Dietrich, 1990). Mount Isa Mines expect to be using this routinely for lead-zinc ore in the near future.

15.2.7 *Core specific gravity measurements*

There are two reasons why one needs to know the specific gravity of core samples:

1. To enable a resource tonnage to be calculated, i.e. one has to be able to convert the volumes of a resource based on drillhole assay intercepts, into tonnes for various ore grades and rocktypes.

2. To enable the average grade of ore to be accurately calculated, particularly where there are significant differences in density of high and low grade ore. Where such differences exist, it is necessary to use the specific gravity to weight individual assays being averaged. To illustrate this point, compare the following differences in weighted averages done with and without specific gravity weighting.

The first of these uses of specific gravity will require a quite accurate average specific gravity for each rocktype or ore grade, or on whatever other basis one wishes to measure tonnage. An error of say 5% in *SG* used, would give an error of 5% in tonnage and hence a similar error in gross revenue, and a much greater variation (maybe 20% or more), in net revenue.

Specific gravity measurements for this purpose must be done in such a manner that an absolute measurement is obtained. This means not just the specific gravity of the solids but of the whole rock with porespace and cavities included. This means that uncrushed core, preferably uncut, must be used, and variations of Archimedes Principle applied. When core is cavernous or oxidised it may be necessary to coat the core in wax or plastic film when immersing in water to obtain volume. When coating is used it is nearly impossible to use all core in any interval for measurements. Broken core in small pieces is difficult to prepare, and will have a different error factor, as the proportion of coating material now becomes a significant proportion of the whole being measured.

To see if coating of core is necessary, test for porosity and its effect (is it permeable?) by placing a length of core in a measuring cylinder of water of similar diameter, and note the change in the displaced volume with time. Where core has a low permeability, (though perhaps a high porosity), and is noncavernous, and the absolute specific gravity is required, then if immersions are rapid, coatings can be avoided as these can introduce errors of their own.

When specific gravity is required for weighting assay intervals, it is only the specific gravity of the solids which is required. In this case the crushed and pulverised core can be used, and after mixing only a small aliquot need be used. As the measurements are to be used in relation to particular assay intervals, these measurements are best made routinely at the laboratory, using a picnometer or using displacement in a flask. Normally, for cost reasons, only the ore intersection and included and adjacent waste will be measured. Where the core is unoxidised and nonpermeable, the results should be the same as those for whole core displacement methods.

If the only reason to do pulp *SG* is to enable *SG* weighted averages, then their relative value is more important than absolute values. Pulp *SG* can however be quite accurate ($\pm$ 5%) for nonpermeable core and so satisfactory for absolute measurements.

The following technique and equipment has been used successfully by the author in a variety of situations and is recommended where large numbers of whole core measurements are to be done on low permeability core. Refer to Figure 67.

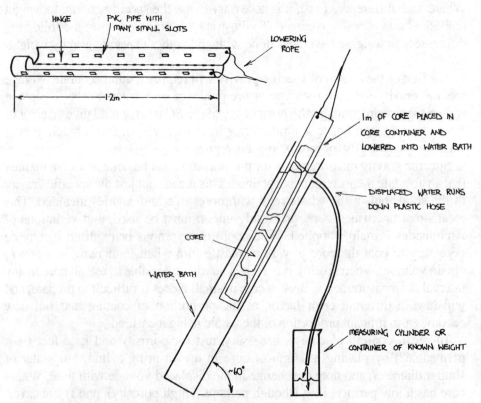

Figure 67. Core volume measurement in field.

Equipment

1. Core container – About 1.2 m of PVC tube (60 mm for *NQ*) with one end blocked and the remainder cut lengthways into halves to be joined with a full length brass hinge. Cut numerous slots or holes into the PVC to allow water to flow through easily. A small piece of rope should be fastened to the open end. As an accessory make a net bag to hold chips of very broken core.

2. Waterbath – About 1.4 m of PVC tube (100 mm for *NQ*) with one end blocked and at about 1.3 m from the blocked end a hole is made to which is fitted a length of 10 mm flexible plastic tube.

3. Measuring cylinder or container of known weight.

4. Scales suitable for measuring in the range 2 to 10 kg, (grocers good scales can often be borrowed).

Procedure

1. Weigh core container ± mesh bag. These weights will be deducted from gross weight later to obtain core weight.

2. Fix waterbath at a steep inclination and fill with water till it begins to overflow from hose.

3. Place an empty measuring cylinder, or alternatively a clean container of known weight under the overflow hose.

4. Slowly slide the empty core container ± the mesh bag into the waterbath, so displacing an equal volume of water into the measuring cylinder or container.

5. Measure volume of container ± mesh bag or weigh the water and so derive the volume. These volumes will be needed to determine the net core volume later.

6. Remove core container from waterbath and top up bath and install empty measuring cylinder under hose.

7. Place core in core container and weigh. The container weight obtained earlier can be subtracted to obtain core weight.

8. Slide the core container into the waterbath and measure volume. The container volume obtained earlier can be subtracted to obtain core volume.

Note: There could be small constant errors because of the scales, surface tension, the rope handle or air bubbles. Because of the large quantity of each sample most of these errors will be insignificant. However, it is a good idea to have checks done at a laboratory.

15.2.8 *Orientated core routines and structural measurements*

Where structural features evident in the drill core require logging and reporting in terms of their inset strike and dip, then orientated core will be required to enable this to be done. Firstly it will need instruction to the drillers so that they can take the cores which can be orientated. The methods by which this can be done have been described in a previous chapter.

Now decisions need to be made on what method will be used and how often an orientated core will be required. As usual this will be a balance of cost, timing and quality. We need therefore to ask how important the ensuing data will be and how easy it will be to obtain? For example if the geometry of a deposit is already well known and very simple and the holes being drilled are infill holes for a resource statement, then one might not rate the acquisition of structural data as all that important. If on the other hand it is first phase, drilling of a geochemical anomaly and there are no surface indications as to the attitude of the indicated mineralization and its hostrock, the structural data will be very important.

If the mineralization is contained within features such as a shearzone (Archean gold deposit), or related to bedding (eg. VMS deposit), then it will be more important than if it is a stockwork such as a porphyry copper deposit.

If the host rock is well bedded or well foliated then almost every time orientated core is drilled, a bedding or foliation plane will be evident and so there can be longer intervals between samples than if the bedding was massive or there was a large percentage of late stage, unrelated dyke rock.

If the rocks are coring well with 100% recovery and have few breaks, then long lengths of core can be fitted back together again to enable it all to be orientated between those pieces of oriented core. Therefore the better the core the further apart the samples need be.

In terms of which method to use, the pencil punch on the basis of simplicity and cost has been found to be the best unless there are special situations such as very broken ground, in which case an orientating core barrel will give the better result, and of course near vertical holes and underground upward holes require other techniques.

So now that we have the core in trays with an occasional (maybe every 15-30 m) piece with a mark by the drillers indicating which way is top or bottom of the core, or the relational mark for an acid tube or survey disk, according to what method was used, we can then proceed.

Firstly these marks should be converted to a mark to indicate top of core as it was drilled. Then the core should be taken from the trays in batches and placed in sequence in a piece of straight channel or angle iron or such. The core should then be fitted together and the top mark transferred along the core with a full line with a black marking pencil. This should then have side arrow marks to indicate downhole direction and should be done so that each piece of core has such a mark.

Of course not all original orientation marks will line up with the next downhole, due to both mechanical and human error. An angular difference of 10° is about normal and any more should be a cause for concern. One common error is where the pencil slides off the point of first impact with little mark but then makes a good mark to one side. I suggest that marks by drillers and others be carefully scrutinised before acceptance.

Now that core has been marked, any structural measurements can be made using a core orienting frame. The most common frame in use in Australia is one made by the staff of James Cook University at Townsville, and there is one made by Eastman (Core Reader) which I have never seen. These allow the core to be positioned in its correct orientation; by that I mean that the core has the same declination and azimuth of the drillhole at that depth, it is rotated so that the top mark is at the top and the arrows point downhole. Structural measurements can now be made in the normal manner as one would make them at an outcrop.

15.2.9 *Core geotechnical logging*

There is no intention here to detail all aspects of geotechnical logging but only to mention those aspects of core preservation, preparation and routine matters which the site geologist can address.

The most important thing to remember is that most geotechnical work should be done on whole core, before any cutting and before extended storage.

Part of the geotechnical logging is of course the core photography to record the appearance of the core breakage pattern so this must be done as soon as possible. With quite broken core this is often best done at the drill site, so that transport induced scrambling and breakage does not cloud the real situation.

The recording of *RQD* (Rock Quality Determination – percentage of core in lengths greater than 10 cm as cored) can be done along with joints per metre and structured data, at an early stage by the site geologist and assistants whilst recording core recovery. In fact it must be done before any cutting for samples. The drillers can also be asked to mark any breaks they make to fit core to the trays so these breaks can be ignored.

The hardness too may change with time and should be recorded early. The recording of this data is best done at metre intervals rather than by core runs as it will make graphical plots much more meaningful. It is usual to set up some local scale of hardness with reference specimens relevant to the rocks being intersected.

The effect of water on some of the more weathered core should also be noted at an early stage. This includes notes on swelling, 'solution', decrepitation, etc. Also any unexplained core losses are best discussed with the drillers as soon as possible as these often have geotechnical implications. The measurement of structural elements and their attitude has been discussed in the preceding section.

APPENDIX 1

Less common core and hole sizes

This is not an exhaustive list but will cover most of those encountered in the field.

Imperial standards and products Name	Core diam. inches (mm)	Hole diam. inches (mm)
XRA, XRT, RWT	0.735 (18.67)	1.175 (29.84)
XRP	0.885 (22.86)	1.290 (32.38)
IEX	0.995 (25.27)	–
EWG, EWM (EX, EXM),	0.845 (21.46)	1.485 (37.72)
EWL, EWF (EXL, EXF),	0.845 (21.46)	1.485 (37.72)
EW D3	0.835 (21.21)	1.485 (37.72)
EWT (EXT), EWK (EXK), TEW	0.905 (22.99)	1.485 (37.72)
EXLT-2	1.062 (26.97)	–
E17	0.968 (24.59)	1.485 (37.72)
EXU	1.040 (26.42)	1.520 (38.61)
AX	1.015 (25.78)	1.852 (34.34)
AXWL	0.937 (23.80)	1.890 (48.00)
AM, AMS	1.032 (26.21)	1.890 (48.00)
AQ, AQU, AV, A18TT	1.062 (26.97)	1.890 (48.00)
AQTK	1.202 (30.53)	–
AMLC	1.062 (26.97)	1.890 (48.00)
AW C3	1.067 (27.10)	1.890 (48.00)
AAWL	1.102 (27.99)	1.890 (48.00)
AW D3	1.136 (28.85)	1.890 (48.00)
A18	1.156 (29.36)	1.890 (48.00)
AMG, AWM (AX, AXM),	1.185 (29.59)	1.890 (48.00)
AWL (AXL), AWF (AXF)	1.185 (29.59)	1.890 (48.00)
AWT (AXT), AWK (AXK)	1.281 (32.54)	1.890 (48.00)
TAW	1.307 (33.20)	1.890 (48.00)
ABQT	1.320 (33.53)	1.890 (48.00)
IAX	1.385 (33.18)	1.890 (48.00)
A17	1.310 (33.27)	1.890 (48.00)
AW34	1.320 (33.53)	1.890 (48.00)
AKU	1.395 (35.43)	1.937 (49.40)
B	1.395 (35.43)	2.105 (53.47)
BM	1.395 (35.43)	2.360 (59.94)

Appendix 1 (continued).

Imperial standards and products Name	Core diam. inches (mm)	Hole diam. inches (mm)
BU	1.610 (40.89)	2.360 (59.94)
BMLC	1.386 (35.20)	2.360 (59.94)
BXWL	1.312 (33.32)	2.360 (59.94)
BQTT, BQ3, BV3	1.320 (33.53)	2.360 (59.94)
BQ 2.32	2.305 (58.55)	2.320 (59.93)
BQWL, BV, BW C3	1.433 (36.40)	2.360 (59.94)
ABWL	1.496 (38.00)	2.360 (59.94)
B18TT	1.500 (38.10)	2.360 (59.94)
BXSM	1.530 (38.86)	2.360 (59.94)
B18	1.565 (39.75)	2.360 (59.94)
BY	1.575 (40.00)	2.360 (59.94)
BW D3, BP D4	1.615 (41.02)	2.360 (59.94)
BWG, BWM (BX, BXM),	1.625 (42.04)	2.360 (59.94)
BWL (BXL), BWF (BXF)	1.625 (42.04)	2.360 (59.94)
BWT (BXT)	1.750 (44.45)	2.360 (59.94)
TBW	1.768 (45.21)	2.360 (59.94)
BW44	1.755 (44.58)	2.360 (59.94)
BXU	1.834 (46.58)	2.375 (60.32)
NXWL	1.718 (43.64)	2.980 (75.69)
NQTT, NQ3, NV3	1.775 (45.08)	2.980 (75.69)
NM, NMS	1.862 (47.21)	2.980 (75.69)
NQWL, NV	1.875 (47.62)	2.980 (75.69)
NW, C3	1.875 (47.62)	2.980 (75.69)
NQ2	1.995 (50.67)	2.980 (75.69)
NXEWL	2.000 (50.80)	2.980 (75.69)
NW 80	2.032 (51.62)	2.980 (75.69)
N18TT, NMLC	2.045 (51.94)	2.980 (75.69)
NW. D3, NW D4	2.060 (52.32)	2.980 (75.69)
NXSM	2.070 (52.58)	2.980 (75.69)
NLC, NXHW	2.085 (52.96)	2.980 (75.69)
N18	2.095 (53.21)	2.980 (75.69)
NWG, NWM (NX, NXM)	2.155 (54.74)	2.980 (75.69)
NX, PAM, NWL, (NXL)	2.155 (54.74)	2.980 (75.69)
NWF (NXF)	2.155 (54.74)	2.980 (75.69)
NQWL BEVEL WALL	2.255 (64.90)	2.980 (75.69)
NDBGM	2.200 (55.88)	–
NWT (NXT)	2.318 (55.89)	2.980 (75.69)
TNW	2.390 (60.71)	2.980 (75.69)
NXU	2.375 (60.32)	2.980 (75.69)
NC	2.735 (69.47)	3.650 (92.71)
NX C 'U'	2.875 (73.02)	3.630 (92.20)
NC	2.695 (68.45)	3.630 (92.20)
NC D3, HW D3, NW D4	2.405 (60.96)	3.650 (92.71)
HQ 3.68	2.605 (66.17)	3.600 (91.44)

Appendix 1 (continued).

Imperial standards and products Name	Core diam. inches (mm)	Hole diam. inches (mm)
HQTT, HQ3, HV3	2.406 (61.11)	3.782 (96.06)
HQWL, HV	2.500 (63.50)	3.782 (96.06)
HMLC	2.500 (63.50)	3.907 (99.24)
HWG, HWF	3.000 (76.20)	3.907 (99.24)
HWT	3.187 (80.95)	3.907 (99.24)
PWF (PXF)	3.627 (92.13)	4.748 (120.60)
PQTT, PQ3	3.270 (83.06)	4.827 (125.15)
PQWL (CP WIRE LINE)	3.345 (84.96)	4.827 (122.61)
SWF (SXF)	4.440 (112.78)	5.748 (145.85)
UWF (UXF)	5.505 (139.83)	6.870 (174.50)
ZWF (ZXF)	6.505 (165.23)	7.870 (199.90)

Metric standards and products Name	Core diam. (mm)	Hole diam. (mm)
36B, 36T	22.0	36.3
46K2, 46K3	24.0	46.3
46Z	28.0	46.3
T46 T2-46	31.44	–
46B, 46T	31.7	46.3
46TT	35.3	46.3
LTK-46	35.3	–
56K2, 56K3	34.0	–
56Z	34.0	56.3
56B, 56T	42.0	56.3
56TT	45.5	56.3
LTK-56	45.2	–
66K2, 66K3	38.0	–
66Z	44.0	66.3
66T6	47.0	–
66B, 66T	52.0	66.3
T6N	57.0	75.7
T6 S-N	47.7	–
76CHD (WL)	43.5	75.7
76×47.7 T6S	47.7	–
76K2, 76K3	48.0	–
76Z	54.0	76.3
76T6	57.0	–
76B, 76T	62.0	76.3
86×57.7 T6S, 86 mm, SK6	57.7	–
86K2, 86K3	58.0	–
86Z	62.0	86.3
86T6	67.0	–
86B, 86T	72.0	86.3
96K2	68.0	–

Appendix 1 (continued).

Metric standards and products Name	Core diam. (mm)	Hole diam. (mm)
T6-H	79.0	99.2
T6S-H	71.7	–
100K2	73.0	–
101TTCHDWL	61.1	101.3
101CHDWL	63.5	101.3
101KO	64.0	–
101CMG	71.0	–
101×71.7 T6S	71.7	–
101K2, 101K3	72.0	–
101Z	75.0	101.3
101T6	79.0	–
101T	83.7	101.3
101B	87.0	101.3
114	85.7	–
114U	92.0	–
116KO	79.0	–
116T6-S, 116K3-5	85.7	–
116K2, 116K3	86.0	–
116Z	90.0	116.3
116T6	93.0	–
116B	102.0	116.3
T6-P	93.0	–
83 mm core	83.0	–
85 mm core	85.0	–
131KO	93.0	–
131T6-S	100.7	–
131K2, 131K3	101.0	–
131Z	105.0	131.3
131T6	108.0	–
131B	117.0	131.3
134CHD	85.0	134.3
141	108.2	–
146KO	108.0	–
146K2, 146K3	116.0	–
146Z	120.0	146.3
146T6, 146T6-S	123.0	146.3
146B	132.0	146.3

Sample drilling contract agreement

The author takes no responsibility for the accuracy of this document and the legalities contained or expressed within it. It is included here as a guide only and anyone who uses part or all should seek separate legal opinion.

Ex Why Zee Mines N.L.

MEMORANDUM OF AGREEMENT
FOR SURFACE DRILLING BETWEEN

Ex Why Zee Mines N.L.

and

Youbeaut Contractors Pty Ltd

AGREEMENT entered into this day   of   19

Ex Why Zee Mines N.L. of 567 Wombat Avenue, Stillton, W.A., Australia, 6999, the party of the first part, hereinafter referred to as the Company.

and

Youbeaut Contractors Pty Ltd of 56 Diidah Street, Bilmedoo, Qld., Australia, 4099, the party of the second part, hereinafter referred to as the contractor.

Whereas the company wishes to have performed certain drilling on mining properties located at Goanna Springs in the Cloncurry District, Queensland, and it is mutually agreed that, in consideration of payment by the Company to the Contractor of the money specified at the time and events mentioned, the Contractor shall carry out and complete the work specified to the entire satisfaction of the Company in every respect.

NOW THEREFORE IT IS WITNESSED:

1. *General*
   a) The whole of the work contemplated under terms of this agreement shall be carried out under the control of the Company's Representative in the particular area under consideration.
   b) All employees of the Contractor engaged in the execution of the work at the location covered by this agreement shall be fully approved by the Company's Representative. Any

employee not so approved shall be replaced and removed from the location by the Contractor within seven days of due notice being served to this effect.

c) No portion of this contract may be sublet by the Contractor without the written approval of the Company.

d) The Contractor shall in all things observe and comply with the rules and regulations of the Company and all other laws and regulations of any other authorities for the time being in force, whereby anything is required to be done or omitted to be done by the Contractor herein.

e) The completion of each drillhole shall include a final and completed clearance by the Contractor of all rubbish, litter, tools, appliances and structures of a temporary nature, and the leaving of the whole area in a clean and tidy condition.

f) All tools, equipment, utensils, and other appliances supplied by the Company for use only during the continuance of the Contract herein, shall be returned at the completion of this contract in good order, fair wear and tear excepted, or failing this their value shall be deducted in the manner provided for deductions herein.

g) The Company shall be entitled to deduct at any time from any payments, any money owing to the Company by the Contractor.

h) The prime purpose of the work specified herein shall be to obtain accurate and complete samples of drill cuttings and/or core, relating to the rocks and minerals penetrated, in their correct sequences, properly related to the depth of the drilling at the places and times required by the Company with the maximum clarity.

2. *Scope*
This agreement is for the drilling programme as specified herein:

Drilling of approximately 9 holes, comprising around 1400 m of precollar percussion drilling (down to 150 m plus) and approximately 1000 m of *NQ* diamond coring.

This programme may be varied subject to the progress results obtained.

3. *Access*
The drilling sites and access thereto will be prepared by the Company or its contractor and maintained by the Company or its contractor during the term of this Agreement, with due regard to the requirements of the authorities.

4. *Depths*
Although the drillhole depths proposed will be subject to variation according to the information required and obtained, the actual depth of each drillhole will be determined by the Company and drilled by agreement.

5. *Reports*
The Contractor shall make a daily record of drilling progress, including the depth of each hole as well as a statement of shut-down time and other activity and all materials used or left in the hole. This record shall be available to the Company whose Representative shall agree the figures, and signify this by signing the Contractor's copy of the record.

6. *Performance*
The Contractor undertakes to have all work agreed to be carried out, performed in a proper and workmanlike manner in accordance with good standard practice appropriate to the drilling industry.

7. *Discipline*
The Contractor shall, at all times, enforce strict discipline and maintain good order among its

employees, and shall not retain on site any unfit person or anyone not skilled in the work assigned to him.

8. *Representation*

The Contractor shall, at all times during the progress of the work, have a responsible person in charge of the work at the location, who shall be authorised to receive instructions and to represent the Contractor for all purposes of this contract.

9. *Sampling*

Drilling shall be conducted so as to produce maximum sample or core recovery, with every reasonable precaution taken to prevent contamination. All samples from the drilling of each drillhole shall be properly identified, and bagged, or in the case of core, placed in correctly labelled core trays by the Contractor, to be presented to the authorised Representative of the Company at the drillhole site, in their correct sequence.

10. *Failure to complete*

The Contractor shall not be entitled to receive any payment for any drillhole if not drilled and sampled to the depth required by the Company, and any such payment for progress shall be recoverable by the Company as a debt due by the Contractor and payable on demand, unless the Contractor can establish to the satisfaction of the Company that the circumstances were beyond the control of the Contractor, in which case he will be paid for the metreage actually drilled and sampled.

11. *Security*

The Contractor shall not give out any information regarding drill results or permit access to drill samples to any person other than the Company's accredited Representatives, except upon specific permission of responsible officials of the Company, and shall take all necessary steps to ensure compliance with this clause by the Contractor's employees.

12. *Injury to persons and property*

The Contractor shall indemnify the Company against any claim for legal liability arising out of personal injury and/or property damage made against or incurred by the Company which results from any act or omission of the Contractor, its servants or agents, but the Contractor shall not be liable to indemnify the Company against any claim or liability which arises solely out of any act or omission of the Company, its servants or agents.

13. *Insurance*

Insurance effected by the Contractor pursuant to clause 13 shall be extended to include the interests of the Company and as such the policy shall indemnify the Company in respect of any action claims demand suit or proceedings.

a) The Contractor warrants that during the terms of the performance of the works, more particularly described herein, it will maintain in effect a policy of insurance to indemnify the Contractor to the full amount of its legal liability to pay compensation under the Western Australia Workers' Compensation and Assistance Act of 1981 as amended or any corresponding enactment in any other State, Territory or Country and the aforementioned policy of insurance will be limited only by such amount as may be set down in any statute and shall in all other respects be unlimited.

b) The Contractor further warrants that it will, during the terms of the performance of the said works, maintain in effect a policy of Public Liability insurance which provides an indemnity to the Contractor to the sum of $1,000,000 against each and every claim that may be made against the Contractor for any legal liability thereby incurred by the Contractor

for:-
  - death or bodily injury (including illness).
  - loss of or damage to property.

14. *Obligations*
Contractor
For the purposes of carrying out the work under this Contract, the Contractor shall, at his own expense:

a) Provide all plant and equipment, including all drilling plant, tools, materials, and all other things necessary for the due and proper completion of the drillholes and collection of the samples therefrom.

b) Accept responsibility for all equipment provided by the Company for the purpose of this Contract.

c) Employ and pay all persons necessary to be employed, and provide meals and accommodation as necessary for those persons.

d) Comply with all regulations and directions as to camping facilities and sanitation as may be required by the properly constituted authorities and the Company.

e) Arrange and complete the transporting of all equipment and personnel as necessary for proper completion of the contract.

f) Accept responsibility for the cost of all materials or equipment loss or damage down hole due to the Contractor's negligence. In addition the Contractor shall not charge the Company for any time losses from attempting to recover such materials or equipment or otherwise incurred as a result of the Contractor's negligence.

Company
For the purpose of enabling the Contractor to carry out the work under this Contract the Company shall, at its own expense:
  - Provide sample bags and core trays at each drillhole site.

15. *Prices and payment*
The Contractor agrees to perform the drilling as required for the performance of this contract in accordance with the attached schedule.

The Contractor agrees that its labour, and all other operating expenses, except as herein provided, shall be at its own expense and for its own account.

16. *Payment for work*
The Company agrees to pay the Contractor in Australian funds, the prices specified on the schedule of charges.

Payment shall be made by the Company within 14 days of the date of invoice. The invoices, in duplicate, shall be rendered twice monthly to:

Ex Why Zee Mines N.L.
567 Wombat Av.,
Stillton W.A.
Australia 6999

17. *Cancellation*
The Company reserves the right to cancel this contract at any time without notice for reason of unsatisfactory sample recovery.

19. *Time*
Time is the essence of the contract.

20. *Determination*
This contract may be regarded as terminated:
   a) After completion by the Contractor of the work specified to the entire satisfaction of the Company and upon final payment by the Company, or
   b) On either party tendering one week's notice of intention to determine, or
   c) On non-observance by one party of the conditions specified and agreed herein, immediately upon notice tendered by the other party not in default.

In witness whereof the parties hereto have executed these presents on the day and year first above written.

Signed on behalf of
Youbeaut Contractors Pty Ltd   . . . .

In the presence of: . . . .

Signed on behalf of
Ex Why Zee Mines N.L.   . . . .

In the presence of: . . . .

SCHEDULE OF CHARGES

Rig – Warman 1000

1. *Mobilisation and demobilisation*
$1500 Total

2. *Reverse circulation percussion drilling*
   0-100 m $20.00/m
100-200 m $22.00/m

3. *Diamond coring*
   0-100 m $60.00/m
100-200 m $63.00/m
200-300 m $66.00/m
300-400 m $69.00/m
400-500 m $72.00/m

4. *Workover (active) rate*

The following, if applicable, will be charged at $140.00/hour:

    4.1 Running, reaming, cutting and pulling casing.

    4.2 Cementing and drilling cement.

    4.3 Mixing mud, mudding holes or otherwise treating lost circulation zones.

    4.4 Changing the rodstring from RC to diamond core and vice versa.

    4.5 Fishing for or recovering lost or broken casing or rods.

5. *Standby rates*

The following, if applicable, will be charged at $100.00/hour:

    5.1 Conditioning hole.

    5.2 Stand-by or other delays caused through fault or request of the Company.

    5.3 Surveying, logging or inserting thin PVC casing for logging purposes.

    5.4 Waiting for cement to set.

6. *Materials*

The following materials will be charged at cost (or agreed depreciated value for used equipment):

    – Casing shoe bits not re-usable.

    – Casing left in hole due to ground condition or at request of company.

    – Down hole tools.

Consumables will be charged at cost plus 10%. These include:

    – Foam.

    – Drilling muds, soluable oils and other additives.

    – Lost circulation sealants.

    – Cement.

7. *Drilling water supply*

By truck (8000 l) $4.50 per km

or

PVC pipe at cost plus hourly rate for laying pipe and setting up pump. . . . . . At work rate

8. *Down-hole surveys*

Camera surveys carried out by the Contractor with the Contractor's own survey equipment
$50.00/shot

Drilling and the environment

A suggested appendix to drilling contracts in New Zealand

The author takes no responsibility for this document and the legalities contained or expressed within. It should be verified by the user.

1. *Access*

Access to an exploration area must not result in any undue disturbance to landowners, their stock or the soil and water resources. Access routes must be established in accordance with the following:

a) Existing tracks, roads and stock races should be used in preference to construction of new routes.

b) Any new access which has to be established should follow a contoured route which minimizes vegetation removal and in accordance with good soil conservation practices and consents.

c) Only a single access route should be used to and from site.

d) Stream beds or rivers should not be used as vehicular access tracks.

e) Stream crossings must be made at constricted points and if crossing already exists it should be used. For small streams and temporary crossings then staked straw or hay bales should be placed just downstream.

f) All gates should be left in the position in which they are found at time of entry.

g) Stock should not be moved from any paddock.

h) Vehicular speed should be reduced to avoid disturbance to stock.

i) Recognize the seasonal nature of farming and avoid maternity paddocks or cropping paddocks.

j) Seek advice from Company environmental staff on soil erosion avoidance.

2. *Drill site preparation*

The drill site must be constructed on the smallest possible area which will allow safe operation of all equipment. This will be done by:

a) Minimize excavation for the level platform.

b) Construct stormwater drainage with appropriate silt controls, and lower the concentration of runoff and hence velocities. Use temporary silt fences, silt traps and ponds as necessary and feasible.

c) Minimize disturbance to vegetation and the ground generally in the surrounds.

d) Where contour drains are required to divert surface water then such drains should be

140

appropriate to the situation. As a guide a sectional area of 0.3 sq.m and longitudinal gradient of 1:40 are suggested.

e) All cutoff drains should be terminated in silt traps and overland flow should be over vegetated surfaces where scouring and rilling will not occur.

f) Topsoil should be stripped and stockpiled.

g) Site preparation will be done under the direction of the Company staff when practical.

3. *Water supply*
This will be arranged to the satisfaction of the Company and the relevant authorities.

a) The extraction and discharge of water from any exploration site is governed by the Water and Soil Conservation Act, 1967, as amended and other statuary approvals.

b) Every contractor will adhere to these regulations, and to enable him to do so he will be given a copy of the conditions which are relevant.

c) Water pumps should be as small as possible.

d) Access to the drill site should be by a single access route.

e) An oil drip tray should be placed under the pump and checked each day.

f) The supply line to the drill site should be laid by hand, with no benching or construction works.

g) No stream works should be undertaken at the pump except for minor works to submerge the intake.

h) No dams will be constructed without the approval of the Company and the relevant authority.

i) Intake should be as close to the bank as possible, and in any case should be less than a third the width of the stream.

j) All intakes should be screened to stop debris and aquatic organisms.

k) Minimum flow requirements as set by the Regional Council must be adhered to.

4. *Water reticulation*

a) All water reticulation around the site must be in closed circuit where possible.

b) Any open channel between the drill site, mud tanks and storage pit must be sized and installed to completely contain the flow.

5. *Water disposal*

a) The ground around the drill collar should be protected from any spillage or flow by the construction of a small open drain to the sludge pit and/or a small bund on the downslope side directing fluids (all fluid which is involved in the activity) to the pit.

b) There will be no direct discharge of excess drilling fluids to an open watercourse.

c) At the completion of drilling all muds will be directed to the sludge pit or taken from site.

d) At site closure all fluids in the sludge pit will be left to drain or evaporate and then the pit will be backfillled and clean fill and topsoil used to cap.

6. *Drilling operations and additives*
Drilling is to be carried out in an orderly and tidy manner with minimum disturbance to landowners, residents and stock. Particular attention is to be paid to control of fluids (loss or make) as follows.

a) If drilling is close to any residence then appropriate noise control measures should be taken.

b) Any water make to the circulating load should be noted and checks made as to how much and from where.

c) The water level in the sludge pit should not rise above a 0.3 m freeboard.

d) If drilling is occurring within 200 m of any watercourse, inspection should be made every 4 hours of operation to check stream water clarity and this logged.

e) There should be a minimum of additives used during drilling. Only those approved by the authorities should be used. Where specified, as conditions of Water Rights, the schedule of approved additives should be set out in the drilling contract.

7. *Rehabilitation and maintenance*
The Company's objective is to rehabilitate sites to a standard as good or better than that which existed prior to drilling.

a) Immediately drilling is completed all equipment is to be removed from site.

b) Silt control devices should be checked and if required, re-established.

c) The site will be recontoured and drainage placed as set out in the Company's specific instructions on this.

d) Topsoil to be replaced and graded to the same contour as the natural ground.

e) The topsoil should be prepared by cultivating the ground either by hand methods or a rotary hoe or such, in two directions across and not down slope.

f) After the seed has been applied, the soil surface will be lightly raked or harrowed to form the final soil surface contour.

g) Areas to be grassed using the hydro-seeding method should be fertilised and seeded according to the Company's standard specifications.

h) The seed mix used will be that as specified by the Company.

i) Vegetation to be re-established such that it is consistent with previous cover and surroundings. Native species should be used but exotic species can be used for immediate soil conservation purposes.

j) Maintenance of the site should be determined after inspections on two occasions at three monthly intervals after site closure.

8. *Landowner agreements and communication*
The following code of practice is intended as a guide to communication with landowners upon whose land exploration activities will be undertaken.

a) A visit to all affected landowners will be made by representatives of the Company and provide the owner with a notice of entry, and will discuss:

Conditions set out in the exploration licence such as gates, firearms, dogs, litter etc.

How the Company's programme can best fit the farm schedule to interfere least.

The extent of the work and number of people, vehicles and plant that will be involved and how to recognise the vehicles at a distance. Identification will be carried out on the first time of entry of all persons who should make themselves known to the landowner.

The landowner will be given a contact person within the Company

b) During and following the work the landowner will be regularly visited by the Company representative to ensure that there are no work related problems or if there are then these are addressed.

c) All landowners within the licence will be kept informed of work progress, even if they are not directly affected by it.

d) All workers (both Company and Contractor employees) will be made aware of the landowner's requirements.

e) The landowner is to be encouraged to make a visit to inspect the works.

f) Upon expiry of the licence, the landowners will be informed of the fact.

9. *Inspection and relations with authorities*

Policing will be open and frank, with no surprises and access should be provided to the relevant regulatory authorities at any time. The following are specific items.

a) Regional Councils often require notification of mobilization to site. The Contractor should notify the Company two working days in advance of the time required by the Council as set out in the Water Right.

b) The Company will inspect the site before drilling commences in conjunction with the Contractor and an agreed procedure will be determined. Further checks by the Company will be done from time to time to ensure the procedures are being implemented.

c) Specific inspections will be made at:
 – Site establishment.
 – At random times during the work.
 – During site closure and rehabilitation.
 – Twice at three monthly intervals following site closure.

d) All communication regarding these guidelines and any of the regulatory authorities will be through the Company.

References

Australian Drilling Industry Training Committee, 1985. *Australian drillers guide*, 2nd edition. Australian Drilling Industry Training Committee.

Barnes, J.F.H., 1989. R.A.B. drilling – secret weapon. In *Resources Review, Oct. 1989*.

Bartrop, S.B. & D.A. Sims, 1990. Mining geology for silver- lead-zinc orebodies at Mt Isa. In *Mine Geologists Conference, Mt Isa, Aus. I.M.M. Oct. 1990*.

Bluck, R.G., 1978. The influence of anisotropic rock properties on diamond drillhole deviation. Unpublished thesis James Cook University of North Queensland.

Clark, L.E. & G.R. Shafto, 1987. Core drilling with Syntax3 polycrystaline diamond. In *Proc. Drillex 87 Conference, Warrickshire, England. I.M.M. 1987*.

Cook, I., 1900. Computer aided core logging at Pasminco Mining, Broken Hill. In *Mine Geologists Conference, Mt Isa, Aus. I.M.M., 1990*.

Cumming, J.D., 1971. *Diamond Drill Handbook*. J.K. Smit & Sons Diamond Products, Toronto, Canada.

Dietrich, P.E., 1990. Accuracy and analysis – the copper bulk analyser at Mt Isa Mines. In *Mine Geologists Conference, Mt Isa, Aus. I.M.M., 1990*.

Downs, R.C., 1990. Geological mapping and logging at Hellyer. In *Mine Geologists Conference, Mt Isa. Aus. I.M.M., Oct. 1990*.

Eggington, H.F. (Ed), 1985. *Australian Drillers Guide*. 2nd ed. Australian Drilling Industry Training Committee, Australia.

He Yizhang, 1987. New diamond drilling techniques in China. In *Proc. Drillex 87 Conference, Warrickshire, England. I.M.M. 1987*.

Longyear, 1989. *Diamond Drill Products Field Manual*, Longyear 1989.

McGregor, K., 1967. *The Drilling of Rock*. C.R. Books, London.

Mills, J.R. (Ed.), 1982. *Drillers Training and Reference Manual*. National Water Well Association of Australia.

Rulisek, P., A. Balcar, K. Cerney, M. Dressler, & J. Vasina, 1987. Studies on the design parameters of impregnated core bits for geological exploratory drilling. In *Proc. Drillex 87 Conference, Warrickshire, England*. I.M.M. 1987.

Schunnesson, H., 1987. Longhole drilling with top-hammer technique- its potential application in thermal heat storage. In *Proc. Drillex 87 Conference, Warrickshire, England*. I.M.M. 1987.

Stevens, D., 1988. Hole deviation and deflection. In *Resource Review, Oct. 1988*.

Suttill, K., 1987. Deflection drilling. In *Engineering and Mining Journal, July 1987*.

Thackray, M., 1984. Slim Hole Drilling Manual. Unpublished by National Lead Co., Baroid Division.

Thorne, M.G., C.A. Jackson, J.K. Bawden & I.P.J. Soper, 1987. In *Proc. Drillex 87 Conference, Warrickshire, England.* I.M.M. 1987.
Yuan Congyu, 1987, Wireline coring coupled with high frequency impactor. In *Proc. Drillex 87 Conference, Warrickshire, England.* I.M.M. 1987.

Subject index